Springer Proceedings in Materials

Volume 16

Springer Proceedings in Materials publishes the latest research in Materials Science and Engineering presented at high standard academic conferences and scientific meetings. It provides a platform for researchers, professionals and students to present their scientific findings and stay up-to-date with the development in Materials Science and Engineering. The scope is multidisciplinary and ranges from fundamental to applied research, including, but not limited to:

- Structural Materials
- Metallic Materials
- Magnetic, Optical and Electronic Materials
- Ceramics, Glass, Composites, Natural Materials
- Biomaterials
- Nanotechnology
- Characterization and Evaluation of Materials
- Energy Materials
- Materials Processing

To submit a proposal or request further information, please contact one of our Springer Publishing Editors according to your affiliation:

European countries: **Mayra Castro** (mayra.castro@springer.com)
India, South Asia and Middle East: **Priya Vyas** (priya.vyas@springer.com)
South Korea: **Smith Chae** (smith.chae@springer.com)
Southeast Asia, Australia and New Zealand: **Ramesh Nath Premnath** (ramesh.premnath@springer.com)
The Americas: **Michael Luby** (michael.luby@springer.com)
China and all the other countries or regions: **Mengchu Huang** (mengchu.huang@springer.com)

This book series is indexed in **SCOPUS** database.

More information about this series at https://link.springer.com/bookseries/16157

Ahmad Osman · Antonia Moropoulou
Editors

Advanced Nondestructive and Structural Techniques for Diagnosis, Redesign and Health Monitoring for the Preservation of Cultural Heritage

Selected work from the TMM-CH 2021

Springer

Editors
Ahmad Osman
Fraunhofer Institute for Nondestructive
Saarland University of Applied Sciences
Saarbrücken, Germany

Antonia Moropoulou
Zografou Campus
National Technical University of Athens
Athens, Greece

ISSN 2662-3161 ISSN 2662-317X (electronic)
Springer Proceedings in Materials
ISBN 978-3-031-03794-8 ISBN 978-3-031-03795-5 (eBook)
https://doi.org/10.1007/978-3-031-03795-5

This Springer imprint is published by the registered company Springer Nature Switzerland AG
The registered company address is: Gewerbestrasse 11, 6330 Cham, Switzerland

Preface

The adoption of innovative scientific methodologies and the implementation of challenging projects that mark future trends in the protection of cultural heritage have initiated a universal and wide-ranging dialogue among scholars, professionals, stakeholders, industry representatives and policymakers, within a holistic approach that fuses competences from the scientific fields of architecture, civil engineering, surveying engineering, materials science and engineering, information technology and archaeology. The advanced nondestructive and structural monitoring and assessment techniques, in conjunction with 3D documentation technologies and analytical testing, support the creation of a transdisciplinary multispectral modelling towards the sustainable preservation of cultural heritage.

These innovative aspects were enhanced and revealed by the 2nd TMM_CH International Conference on "Transdisciplinary Multispectral Modelling and Cooperation for the Preservation of Cultural Heritage: Recapturing the World in Crisis through Culture", during 13–15 December 2021, in Athens, Greece.

The conference was organized by the National Technical University of Athens in cooperation with the Technical Chamber of Greece, under the patronage of H.E. the President of the Hellenic Republic, Ms Katerina Sakellaropoulou, inaugurated by H.E. the Vice President of the Government of the Hellenic Republic, Mr Panagiotis Pikrammenos, with benedictions bestowed by His All Holiness, Ecumenical Patriarch, Bartholomew I of Constantinople, and His Beatitude Archbishop Hieronymus II of Athens and All Greece.

Distinguished scientists and representatives of the National Geographic Society, the Cultural Heritage Finance Alliance, the International Council of Monuments and Sites ICOMOS, the Organization of World Heritage Cities OWHC, the European Society for Engineering Education SEFI, the European Construction Technology Platform ECTP, the International Federation of Surveyors FIG, the International Committee CIPA Heritage Documentation, the World Monuments Fund, AHEPA Hellas, UNESCO Chairs on "Digital Cultural Heritage" and "Culture, Tourism, Development", and other major International and European Organizations, Associations, Networks, Universities and Research Centers in the field of cultural heritage preservation, participated in the International Steering and Scientific Committees.

The 2nd TMM_CH Conference was highly expected and attracted researchers from all over the world. It was held with great success, despite the pandemic, with the physical presence of 150 delegates and online attendance of 500 delegates in real time. A total of 310 contributions were submitted, and 124 papers were accepted for oral presentation and publication by 377 authors from 33 countries, after peer review from three reviewers in average per paper and consequent revisions, with a rate of acceptance equivalent to a 40%. Accepted papers were published in distinguished scientific journals and proceeding volumes. Sixteen papers are presented in this book. The editors would like to express their sincerest appreciation to all of the reviewers and authors who contributed in maintaining the high standards necessary for the implementation of the successful 2nd TMM_CH Conference, as well as for the preparation of this book.

Within this book, the benefits of Nondestructive Testing (NDT) and their applications are demonstrated on several diverse and important cultural heritage assets, ranging from artefacts, to historic buildings, and to historic areas around the world. NDT methods are today essential tools for the characterization of an artefact's or a cultural heritage asset's constituent materials and for the study of their manufacturing process, thus, improving our understanding of cultural heritage, as well as of their materials. These are prerequisites for a thorough assessment of the preservation state of a heritage asset and for the design of effective, compatible and sustainable interventions to enhance its resilience against the prevailing risks and preserve its values. This book should be beneficial for scientists, engineers, architects, archaeologists, conservators, geologists, art historians and interested stakeholders and public, as a concise source of recent scientific advancements, nondestructive techniques and technologies for monitoring, documenting and preserving, in a holistic approach, the world cultural heritage.

May 2022

Ahmad Osman
Antonia Moropoulou
Kyriakos Lampropoulos

Contents

About the Editors

Prof. Dr.-Ing. Ahmad Osman - Fraunhofer Institute for Nondestructive Testing - IZFP, Department Manager Algorithms/Signal- und Data Processing, Head of the Applied AI group AutomaTiQ, Member of the International Steering and Scientific Committees of the 2nd International Conference TMM-CH "Transdisciplinary Multispectral Modelling and Cooperation for the Preservation of Cultural Heritage Recapturing the World in Crisis through Culture".

Dr. Antonia Moropoulou, Professor Emeritus - National Technical University of Athens, President of the International Steering and Scientific Committees of the 2nd International Conference TMM-CH "Transdisciplinary Multispectral Modelling and Cooperation for the Preservation of Cultural Heritage Recapturing the World in Crisis through Culture".

Dr. Kyriakos Lampropoulos, Educational and Research Staff - National Technical University of Athens, Member of the International Scientific Committee of the 2nd International Conference TMM-CH "Transdisciplinary Multispectral Modelling and Cooperation for the Preservation of Cultural Heritage Recapturing the World in Crisis through Culture".

Advanced Nondestructive and Structural Techniques for Diagnosis, Redesign and Health Monitoring

Unraveling the Materials and Techniques of Post-Byzantine Icons from Epirus, Greece

Anthoula Poulia[1(✉)], Konstantina Sougioultzi[2], Ioannis Liougkos[3], and Alkiviadis Paipetis[2]

[1] Department of Physics, University of Oslo, 0371 Oslo, Norway
anthoula.poulia@smn.uio.no
[2] Department of Materials Science and Engineering, University of Ioannina, 45110 Ioannina, Greece
[3] Art Restoration, Conservation of Works of Art, 45500 Ioannina, Greece

Abstract. This work is dedicated to the investigation of several Post-Byzantine icons from the region of Epirus in Greece, utilizing physicochemical methods. Nondestructive analyses of Fourier – Transform Infrared and Raman spectroscopy were mainly used as suitable techniques for providing information on materials, practices, and alterations, information which is essential for art historians, conservators, and restorers. The techniques used, were selected based on their complementarity to reconstruct the Painter's palette, and to investigate the painting techniques used through the matching of the pigments. The dating of the pigments also revealed interesting findings about the presence of icons that had been undergone older conservation processes. The surface of selected icons was also investigated utilizing Scanning Electron Microscopy, combined with quantification analysis, and additional degradation signs where revealed. The examined icons, covering a time frame from the beginning of the 17th century to the first half of the 19th century, can lead us to draw conclusions contributing to the formation of a mosaic of data about the type of pigments, the organic layers and the techniques selected, in order to finally discover the differences that may exist over time, in the analyzed artistic creations in the region of Epirus, Greece.

Keywords: Post-Byzantine iconography · Epirus · Greece · Pigments · FT-IR spectroscopy · Raman spectroscopy · SEM/EDX · Chioniadite painters · Kapesovites painters · Temple of Vlacherna · Temple of Parigoritissa

1 Introduction

Portable religious paintings made and used in the context of Orthodox Christian worship are called "portable icons" or simply "icons". The introduction of icon painting dating back to the early Christian times, is believed to have peaked - in terms of the of the works' quality characteristics - during the Byzantine period. However, and despite Ottoman and Venetian domination, the production of portable icons continued in Greece and during the post-Byzantine period (1453–1830), with the production of high quality artworks.

A. Osman and A. Moropoulou (Eds.): *Advanced Nondestructive and Structural Techniques for Diagnosis, Redesign and Health Monitoring for the Preservation of Cultural Heritage*, SPM 16, pp. 3–15, 2022.
https://doi.org/10.1007/978-3-031-03795-5_1

As such, during the 15th–17th century, the art of religious painting cherishes its highly conservative character. The preservation condition of Post-Byzantine paintings largely reflects a long environmentally driven decay process. To this end, temples in the Epirus regional district of North-Western Greece, containing icons painted during this period, were selected for this study.

In the region of Epirus, the art of monumental painting was boosted during the sixteenth century [1]. The process that Painters selected to create an icon, usually followed a well-defined protocol [2]. However, in relation to the Post-Byzantine art of panel painting, there was a transitional period (16th till early 19th century). In this period, because of the western European influences, the artists started experimenting with materials and styles [3]. Identification of the coloring materials contained in a work of art is therefore important for the development of effective and appropriate conservation strategies. Furthermore, this strategy provides historical information, necessary for the artwork's documentation [4].

Under this frame, this study aims to determine (a) the current state of preservation, (b) the scale of prior damage, (c) the number and nature of the pigments used, and (d) the characteristics of the structural materials of the examined icons derived from the region of Epirus, in Greece.

2 Experimental

2.1 Points of Interest and Sampling Scheme

The studied icons were derived from the region of Epirus, in Greece. In particular, three portable icons were examined from the region of Ioannina, that were painted by some prominent, seasoned and multifarious painters of Epirus.

The first icon is The Virgin Odigitria Avel, from the Holy Monastery of Avel (Pogoni, Bissani) (Fig. 1a). It dates back to around 1800 and is attributed to a Chioniadite painter since the frescoes were inscribed with the names of the chioniadites Constantine and Michael Michael. The Holy Monastery of Avel is a single-aisle basilica with a dome and a semicircular niche in its north and south side.

Fig. 1. a) The icon of The Virgin Odigitria Avel; b) The icon and The Virgin Odigitria and c) The icon of Saint Anastasios of Persia from Ioannina, Epirus region.

The second icon, The Virgin Odigitria, in the same iconographic style as the first icon, is dated around the second half of the 18th century and it is painted by a Kapesovite painter (Fig. 1b). The Virgin Odigitria is an iconographic type of the Virgin Mary in which she is depicted standing, holding the infant Christ and slightly turned to the left.

The third icon to be examined is that of Saint Anastasios of Persia, which dates to the first half of the 19th century and is attributed to the Painter Theodosius of Ioannina (Fig. 1c).

The next six of the hereby presented portable icons, come from the city of Arta in Epirus, Greece, which was known as the first capital of the Despotate of Epirus (1205–1337/40, 1430–49). The icons belong to two temples that are of special cultural value and thus deserve a brief introduction.

Firstly, the temple of Vlacherna is located in the village of Vlacherna, opposite of Arta city. It was named after the famous Blessed Virgin Mary in Constantinople. In the middle of the 13th century three domes were added, while at the end of the 13th century, the western narthex and the lateral arcades were added. Inside the temple, some parts of the original marble iconostasis are preserved. The current appearance of the iconostasis is the result of additions and its different parts date back to the 17th–19th century. On the north and south wall of the main temple there are two box-shaped tombs. The frescoes of the church belong to two eras. The frescoes of the Sacred Step and the main church date from the middle of the 13th century, while the wall paintings of the narthex are dated in the late 13th century.

For the Vlacherna temple, the experimental investigation was focused on three icons illustrated in Fig. 2. These were: 1) The icon of Christ enthroned, dated in 1658 (Fig. 2a), 2) The icon of the Three Holy Hierarchs, dated in 1867 and painted by Chioniadite painter Michael Constantine (Fig. 2b) and 3) The icon of Christ, dated in 1871 (Fig. 2c).

Fig. 2. a) Icon of Christ enthroned (1658); b) The Three Holy Hierarchs (1867) and c) The icon of Christ (1871) in Vlacherna temple of Arta, Greece.

The second temple is in Arta. The Parigoritissa temple was built in the 13th century, and it is located near the center of Arta city. Known today as The Virgin Parigoritissa, the temple is dedicated to the Annunciation and is one of the most important Byzantine monuments in the region of Epirus. The temple belongs to the octagonal cruciform type, while outside it is a cubic rectangular building. In the interior, it stands out for its lightness and elegance. In the built temple, which replaced the original marble one, there are distinguished successive layers of frescoes, dating from the 16th to the 20th century. In 2014, restoration actions took place in the temple.

In Parigoritissa, two large dimensional icons, to the left and right of the wood-cut altar door, were of great interest. The first one was the icon of the Virgin Mary Odigitria, dated in the 17th century (Fig. 3a) and the second one was the icon of Christ in the type of the High Priest, dated according to the inscription, in 1758 (Fig. 3b). Additional investigations were also performed in the wood-cut altar door of the temple (Fig. 3c).

Fig. 3. a) The icon of the Virgin Mary Odigitria in the iconostasis; b) The icon of Christ in the type of the High Priest in the iconostasis and c) The wood-cut altar door of Parigoritissa temple.

Regarding the sampling scheme of all icons, the basic points of interest were categorized as follows: a) Faces (e.g. eyes, light-colored areas and shady sides), b) Cloaks (depending on the color changes) and c) Additional points of interest (e.g. flesh areas (such as the hands or the neck) and the halos).

2.2 Methods

Infrared spectra were recorded on an Agilent 4300 Handheld portable FT-IR interferometer in the 4000–700 cm^{-1} region, with 64 scans and a resolution of 4 cm^{-1}. Raman spectra were recorded with a Labram HR - Horiba scientific system. The 514.5 nm line of an argon ion laser operating at 1.5 mW at the focal plane was employed for the Raman excitation. An optical microscope served as the collector of the Raman scattering equipped with a $50 \times$ long working distance objective. Raman spectra in the range of 0–3500 cm^{-1} were collected. Finally, specific samples were investigated utilizing Scanning Electron Microscopy, accompanied by Energy Dispersive Spectroscopy (JEOL 6510 LV SEM/EDX) for both microstructural and quantification analysis.

3 Results and Discussion

3.1 Preparation and Pigments Identification Utilizing FT-IR Spectroscopy

Infrared analysis was carried out on areas representative of the palette of each icon. The pigments identified in the painting are listed in the Appendix.

The exact interpretation of the results from the analysis of binding materials may be difficult due to the complex nature of the binders used. However, the following observations were clearly made: All the spectra present a common intense vibration at 1750 cm^{-1}, that can be attributed to the presence of egg yolk. Additionally, wide and intense absorbances occurring around 3000 cm^{-1}, which are ascribed to the O-H stretching modes and coupled with the intense band at 1750 cm^{-1}, suggest the presence of plant resins.

In particular, for the three icons that originated from Ioannina region, the presence of linseed oil is observed, a material that could have been used in the preparation, as a binder or as a coating varnish. In the Icon of The Virgin Odigitria (Fig. 1b) as well as the Icon of Anastasios of Persia (Fig. 1c), occurrence of calcium carbonate was observed, a white pigment, in almost all measurements. This may reveal its use in the preparation process. White materials consisting of calcium carbonate ($CaCO^3$) have occasionally been used as pigment cements. These include chalk, shell white, as well as materials prepared by calcium carbonate treatment (e.g., calcium hydroxide-Ca $(OH)_2$). On the contrary, regarding the Icon of The Virgin Odigitria Avel (Fig. 1a), no such pigments or materials were found.

Regarding the pigments used on the flesh areas (faces and hands), in The Virgin Odigitria Avel (Fig. 1a) a mixture of malachite and yellow ochre was found. Additionally, $BaSO_4$ was present, probably used in later years for aesthetic restoration. Malachite (basic copper carbonate - $Cu_2(OH)2(CO_3)$) is a secondary copper mineral and is found in the upper oxidation zones of copper deposits. It is usually found along with azurite (but more often in greater quantities than the latter), as well as with other copper, calcium, manganese, etc. The darkest points in the faces (shading), proved to be a mixture of malachite and ochre. In the red areas of The Virgin Odigitria Avel, i.e. in the cloak (maphorium) of the Virgin Mary and in the red tunic of Christ, the pigments of cinnabar and the red of lead (Pb_3O_4) were observed. Pigments in which complementary use was widespread since cinnabar, natural or synthetic, remained at high cost and was adulterated with minium [5]. The same red pigments were used in the other two icons from Ioannina. In addition, in the red areas of the other icons, two more pigments were observed; those of hematite and vermilion, a reddish pigment derived from crushed cinnabar (HgS).

In the blue areas of The Virgin Odigitria Avel, a mixture of the following pigments appear azurite and ultramarine in addition to indigo blue dye. Azurite, as well as Indigo blue, are substances with very frequent use in Byzantine and post-Byzantine icons and there is a large volume of literature that supporting their extensive use [6–9]. Azurite (basic copper carbonate - $Cu_3(OH)_2(CO_3)_2$) is therefore considered to be the most important blue pigment in European painting from the 15th to the mid-18th century. Ultramarine is present in natural and synthetic forms. The natural form of the pigment is also known as lazurite. Lazurite had gained the preference of painters during the Byzantine period, due to its intense blue color, but during the years of the Post-Byzantine period it was used to a lesser extent, due to the poverty of the time, and for this reason, it was often adulterated with azurite. Lazurite-derived cement is also undoubtedly the most valued natural pigment both because of its intrinsic - due to the scarcity of raw material - value and because of its high-quality color. In the green cloak of Agios Anastasios of Persia (Fig. 1c), we observed the dye of Indigo blue combined with pigments like azurite, cobalt blue, and sienna. Cobalt blue, was produced at the start of the 19th century, and can be attributed to older restoration.

Traces of kaolin were also present, which may betray ochre yellow. Traces of dried oil, specifically poppy oil, were also found on the cloaks. Poppy oil is produced from the seeds of the poppy (Papaver Somniferum) and contains smaller amounts of unsaturated fatty acids than linseed oil. There may be two reasons for its presence, initially, the use of poppy oil with traces of egg yolk was quite common to be used for yellow and blue areas because it does not affect the color hue over time [10]. Visually, gold was also observed on the Saint's cloaks, leading us to a second explanation for the presence of poppy oil, in that it was used as an oil mordant for the gold leaf.

In the icon of The Virgin Odigitria (Fig. 1b), the pigments that were observed on the complexion were malachite, ochre, white zinc, and the organic dye of alizarin yellow. In the darker areas, stronger peaks for the yellow pigments were observed, while in the lighter ones the peak of the white zinc was particularly noticeable. In The Virgin Odigitria, the carbon black pigment was observed in the maphorium, probably placed as a base under intense painting layers for a better optical effect to be achieved.

Finally, for the Halo areas, in The Virgin Odigitria, traces of white lead, malachite and yellow ochre were found. Ochre, as well as lead white, are commonly used in gold image areas. In the context of painting on wooden supports, the most important white pigment from antiquity until the 19th century was undoubtedly the artificial white of lead. This popular and toxic substance has been widely used as white, but also as a means of achieving lighter tones of other pigments, either by mixing with them or in the form of white substrates/signatures. Malachite can be attributed to the background of the painting layer. The same pigments were also found in the icon of Agios Anastasios of Persia with the addition of Alizarin yellow. In particular, in the icon of Saint Anastasios of Persia's complexion area (Fig. 1c), pigments such as ochre yellow and malachite were also identified.

Moving on to the Vlacherna icons, regarding the icon of Christ enthroned (Fig. 2a), of great interest are the gold areas (e.g. in the cloak stripes) where at $\sim$ 3560, 3120, and 1630 cm^{-1} peaks were recorded for brown ochre ($Fe_2O_3 \cdot nH_2O$, clay, silica). Additionally, peaks at $\sim$ 3380, 1400 and 1020 cm^{-1} correspond to brown ($Fe_2O_3 + MnO_2 +$ clay), while peaks at $\sim$ 800, 900, and 780 cm^{-1} are related to the presence of the fawn ochre.

In the Three Holy Hierarchs icon of Vlacherna (Fig. 2b) of great interest were the violet stripes corresponded to hematite and Egyptian blue, while the white line on these pieces revealed the presence of Pb, due to the possible use of cerussite ($PbCO_3$), or white lead ($2PbCO_3 \cdot Pb(OH)_2$). For the rest of the pigments (e.g. red, blue, white), the same matches were done as mentioned before.

Finally, in the icon of Christ (Fig. 2c) the flesh tone regions were a mixture of yellow ochre and malachite pigment. Additionally, the presence of barium sulfate was identified. In the red cloaks, the peaks at $\sim$ 1550 cm^{-1}, $\sim$ 1400 cm^{-1} and 1300 cm^{-1} were matched with the hematite pigment, the lead red (Pb_3O) and the cinnabar, respectively. The blue areas were matched well with the azurite pigment at $\sim$ 1450 cm^{-1}, the blue Indigo dye at $\sim$ 1050 and 1230 cm^{-1}, the cobalt blue pigment at $\sim$ 990 cm^{-1} and the ultramarine blue (lazurite) at $\sim$ 850 and 1300 cm^{-1}, respectively. The white areas revealed the use of lead white, due to the peak at 1700 cm^{-1}, the titanium white at ~ 3350 cm^{-1} and barium sulfate at ~ 1000 and 800 cm^{-1}, respectively.

In Parigoritissa temple, the icon of the Virgin Mary in the iconostasis (Fig. 3a) and the light-colored complexion areas were matched with the following peaks: yellow ochre (α-FeOOH + kaolin) at ~ 3690 cm^{-1}, titanium white at ~3350 cm^{-1}, presence of chalk at ~ 900 cm^{-1}, ochre yellow at ~1400 cm^{-1}, malachite pigment at ~ 1200 cm^{-1} and yellow Alizarin dye at ~ 1070 and 780 cm^{-1}. Additionally, several points corresponding to different shades of the red color matched well with the following peaks: the natural red earth (Fe_2O_3 + kaolin) at ~ 3620 and 795 cm^{-1}, the cadmium red at ~ 1550 cm^{-1}, and the red lead (Pb_3O_4). at ~ 1400 cm^{-1}. Lead red, so-called minium, is an alternative to cinnabar, which artisans use. It is a naturally occurring lead tetroxide (Pb_3O_4), found as a secondary corrosion product of galenite minerals (lead sulphide-PbS) and cerusite (lead-$PbCO_3$ carbonate).

Regarding the icon of Christ in the type of the High Priest in the iconostasis of Parigoritissa (Fig. 3b), the gold background of the icon that covers a big area of interest was matched with the presence of peaks corresponding to the following: at ~ 3340, 3120 cm^{-1} and 1680 cm^{-1} gold ochre and light ochre (α-FeOOH, gypsum), at 2160 cm^{-1} spinel yellow, at ~ 1630, and 1550 cm^{-1} satin ochre (α-FeOOH + some minerals) and finally at ~ 1050, 980 and 800 cm^{-1} cadmium yellow (CdS) medium (Kremer Pigment). The complexion was found to be a combination of yellow ochre, titanium white and chalk, while traces of malachite pigment and yellow Alizarin were also detected there.

Finally, the wood-cut altar door of the Parigoritissa temple (Fig. 3c) matches its red areas with the natural red earth (Fe_2O_3 + kaolin), the Cadmium red (CdS + CdSe), and the red lead (Pb_3O_4). The green areas correspond to malachite ($CuCO_3{\cdot}Cu(OH)_2$) at ~ 3400, 1490, and 1100 cm^{-1} and probably cobalt green (CoO·ZnO) at ~2400, and 1170 cm^{-1}. In the blue areas the presence of azurite ($2CuCO_3{\cdot}Cu(OH)_2$) at ~ 2230 cm^{-1}, 1230 cm^{-1}, and 800 cm^{-1} correspond to Ultramarine blue while cobalt blue ($CoO{\cdot}Al_2O_3$) was matched at ~ 1160 cm^{-1}. The white areas that match well with the peaks at ~ 3535 cm^{-1}, 2910 cm^{-1} 2850, and 1730 cm^{-1} correspond to lead white ($2PbCO_3{\cdot}Pb(OH)_2$), the peak at 3350 cm^{-1} correspond to titanium white (TiO_2) and the peaks at ~1070 cm^{-1} and 810 cm^{-1} correspond to the presence of BaSO4 respectively. The presence of titanium white, BaSO4 and cobalt blue could be attributed to color completion during older maintenance.

3.2 Pigments Identification Utilizing Raman Spectroscopy

Additional selected Raman investigations were also performed on specific icons to verify the previous FT-IR results. The selection of the icons was limited by the avaiability to be transferred into the Raman laboratory for testing, as well by their size for fitting on the optical table of the instrument.

As such, regarding the icon of Saint Anastasios of Persia from Ioannina (Fig. 1c), Raman analysis in the area of the light-colored region of the face, indicated the presence of a lead-based yellow at 252.1 cm^{-1} and ochre yellow at 340.7 cm^{-1} as depicted in the corresponding data of Fig. 4.

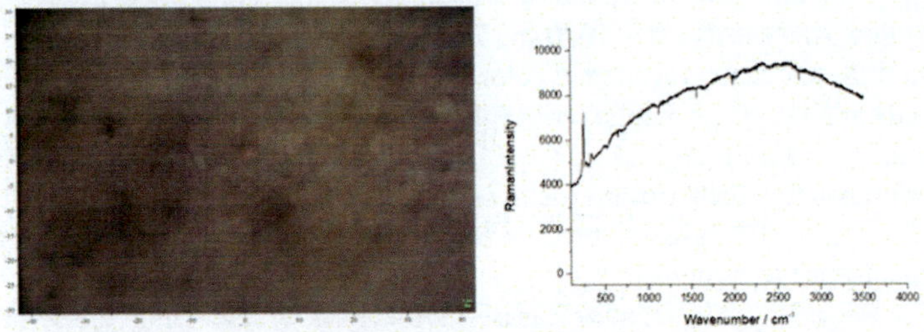

Fig. 4. Raman data for Saint Anastasios from Persia (light colored area of the face).

Finally, Raman spectroscopy applied in the icon of The Virgin Odigitria from Ioannina (Fig. 1b) at the light-colored area of the neck and at the red cloak (Fig. 5a and b respectively), revealed in the first case the presence of a lead-based yellow at 256.1 cm^{-1} and 287.4 cm^{-1} and in the second case the presence of cinnabar at 256.4 and 345.7 cm^{-1} as well as the chalk at 288.2 cm^{-1}. These results further support the FT-IR data, targeting on pigments identification by utilizing combinatorial physicochemical methods.

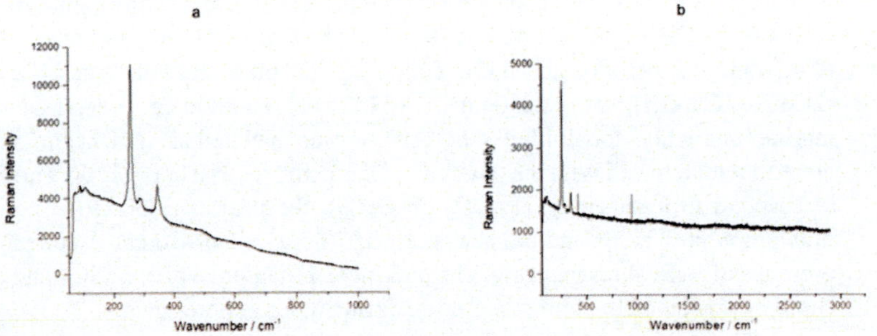

Fig. 5. Raman data for The Virgin Odigitria a) at the light-colored area of the neck) and b) at the red cloak.

3.3 Surface Examination and Elemental Quantification Analysis Utilizing SEM/EDX

Finally, the utilization of SEM/EDX in the icon of The Virgin Odigitria Avel, targeted on the neck area, revealed that the surface of the portable icon bears some small signs of degradation, which are reflected in the form of micro-cracks and micro-bumps (Fig. 6).

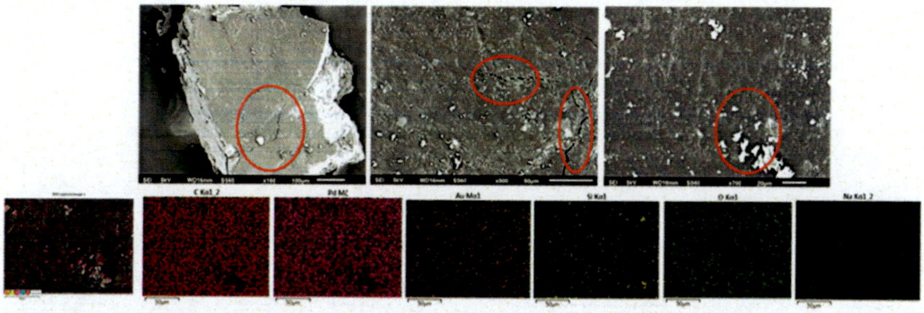

Fig. 6. SEM/EDX data of the Virgin Odigitria Avel icon, derived from the neck area.

Quantitative mapping of EDX Analysis on the same area further verified the presence of oxygen to a greater extent, as well as Na and Si. As a result of the sputtering process, the Pd and Au elements were also detected.

4 Conclusions

This work presented comparative results from two basic spectroscopy techniques (FT-IR and Raman) in the investigation of icons from the region of Epirus in Greece. The observed peaks from the relevant diagrams at each target point were identified after studying the relevant literature. The combination of the techniques was also a beneficial tool. Interesting data was discovered on the variety of pigments used while the dating of the pigments showed that some of them were produced en masse after the first quarter of the 19th century. Thus, hinting at older maintenance. Finally, utilizing microscopy in selected samples and areas also demonstrated the degradation of the examined icons and promoted the quantitative analysis of the relevant spots. It should be pointed out, however, that these techniques alone cannot give unequivocal answers to the problem of paint identity. Evidently, the use of additional analysis techniques such as XRF, UV-VIS, and/or XPS for inorganic substances and liquid chromatography for organic components, could help in more indubitable pigment identification.

Acknowledgements. The present work was funded by the Region of Epirus under the framework of the Interreg Greece-Italy 2014-2020 Program entitled "CI-NOVATEC - Customer Intelligence for inNOVAtive Tourism Ecosystems".

Appendix

Icon		Points of interest	Pigments identified with FT-IR										
			Ochre Yellow	Malachite	Hematite	Lead red	Cinnabar	Azurite	Blue indigo	Co Yellow	Ultramarine	Lead Based Yellow	$BaSO_4$
Region of Ioannina	Icon of The Virgin Odigitria And		x	x									x
			x	x									x
			x	x						x			x
						x	x						
						x	x						
								x	x		x		

Icon		Points of interest	Pigments identified with FT-IR										
			Ochre yellow	Alizarin Yellow	Hematite	Lead red	Cinnabar	Ivory black	Ultramarine	Indigo Blue	Lead white	Azurite	Malachite
Region of Ioannina	Icon of The Virgin Odigitria		x	x									x
			x	x									x
											x		
					x	x	x	x					
					x	x	x				x		
									x	x		x	

Icon		Points of interest	Pigments identified with FT-IR										
			$BaSO_4$	Ochre	Hematite	Cinnabar	Lead red	Cobalt Yellow	Indigo Blue	Azurite	Lead white	Yellow ochre	Malachite
Region of Ioannina	The Icon of Saint Anastasios of Persia		x	x								x	x
			x	x				x				x	x
			x	x							x		x
				x					x	x			
					x	x	x						
					x	x	x						

Fig. 7. Pigments examined by FT-IR spectroscopy in the icons derived from the region of Ioannina.

Icon		Points of interest	Pigments identified with FT-IR													
			Ochre yellow	Malachite	Hematite	Lead red	Cinnabar	Ivory black	Bone black	Azurite	Blue Indigo	Co blue	Ultramarine	Lead white	Titanium white	Yellow Alizarin
Region of Arta	Icon of Christ enthroned_Vlacherna		x	x												
					x	x	x									
								x	x							
										x		x	x			
														x	x	
										x	x	x				
			x	x												x

Icon		Points of interest	Pigments identified with FT-IR												
			Ochre yellow	Yellow Alizarin	Hematite	Lead red	Cinnabar	Ivory black	Bone black	Co green	Viridian green	Lead white	Titanium white	Azurite	Malachite
Region of Arta	The Three Holy Hierarchs_Vlacherna		x	x											
					x	x	x								
								x	x						
										x	x				
												x	x		
														x	x

Icon		Points of interest	Pigments identified with FT-IR													
			Gold ochre	Stroudium yellow	Hematite	Cinnabar	Lead red	Brown ochre	Brown	Fawn ochre	Titanium white	Zinc white	Lead white	Yellow ochre	Malachite	Yellow Alizarin
Region of Arta	The Three Holy Hierarchs_Vlacherna		x	x												
					x	x	x									
								x	x	x						
											x	x	x			
											x			x	x	x

Fig. 8. Pigments examined by FT-IR spectroscopy in the icons derived from the region of Arta, temple of Vlacherna.

Icon	Points of interest	Gold ochre	Satin ochre	Cadmium yellow	Malachite	Raw umber greenish	Verdigris green	Cadmium red	Red lead	Red earth	Yellow ochre	Titanium white
Region of Arta – Virgin Mary Odigitria_Parigoritissa		x	x	x								
					x	x	x					
	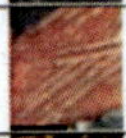							x	x	x		
					x						x	x

Icon	Points of interest	Gold ochre	Spinel yellow	Satin ochre	Cadmium yellow	Azurite	Malachite	Cinnabar	Cadmium red	Red lead	Yellow ochre	Titanium white	Yellow Alizarin
Region of Arta – Christ in the type of the High Priest_Parigoritissa		x	x	x	x								
						x	x						
								x	x	x			
							x				x	x	x

Icon	Points of interest	Malachite	Verdigris green	Raw umber greenish	Yellow ochre	Satin ochre	Azurite	Ultramarine	Cobalt blue	Cadmium red	Red lead
Region of Arta – Wood-cut altar door_Parigoritissa		x	x	x							
	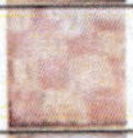				x	x					
							x	x	x		
										x	x

(Each table is headed "Pigments identified with FT-IR" over the pigment columns.)

Fig. 9. Pigments examined by FT-IR spectroscopy in the icons derived from the region of Arta, temple of Parigoritissa.

References

1. Acheimastou-Potamianou, M.: Questions of 16th-century monumental paintings: the provincial epirote school. The Deltion of the Christian Archaeological Society (Deltion) **16**, 13–32 (1992). (in Greek)
2. Hermeneia of the art of Painting. In: Papadopoulos-Kerameus, A., (ed.) Petersburg (1900). [Available online] https://anemi.lib.uoc.gr/metadata/2/c/a/metadata-141-0000170.tkl
3. Vokotopoulos, P.A.: Byzantine Icons. Ekdotiki Athinon, Athens (1995)
4. Karapanagiotis, I., Minopoulou, E., Valianou, L., Daniila, S., Chryssoulakis, Y.: Investigation of the colourants used in icons of the Cretan school of iconography. Anal. Chim. Acta **647**, 231–242 (2009)
5. Vandenabeele, P., Moens, L.: Pigment Identification in Illuminated Manuscripts. Elsevier, Amsterdam (2004)
6. Stasinopoulos, S.: The restoration of the enthroned Virgin. Archeology Arts **22**, 53–58 (1987)
7. Drandaki, A., Branopoulou, L., Kalligas, A.: The Study of the icons signed by Emmanuel Lampardos at the Benaki museum with the contribution of the technical examination. Bull. Christ. Archaeol. Soc. **21**, 189–220 (2000)
8. Aloupi, E., Paschalis, V., Stasinopoulos, S., Tornari, V., Anglos, D., Karydas, A.G., Gionis, V., Chrysikos, G.: Examination, analysis and documentation of the Baptism of Christ by Domenikos Theotokopoulos with non-destructive physicochemical techniques. Benaki Museum **5**, 87–114 (2005)
9. Karapanagiotis, I., Lampakis, D., Konstanta, A., Farmakalidis, H.: Identification of colourants in icons of the Cretan school of iconography using Raman spectroscopy and liquid chromatography. J. Archaeol. Sci. **40**, 1471–1478 (2013)
10. Fitzhugh, E.W.: Red Lead and Minium, in 'Artist's Pigments: a handbook of their history and characteristics', 1, R. L. Feller, National Gallery of Art, Cambridge University Press, Cambridge (1986)

Nondestructive Assessment of Post-Byzantine Icon

G. T. Varfi[1](✉), A. Asvestas[1], D. A. Exarchos[1], S. Farmaki[1], G. Mastrotheodoros[2,3], D. F. Anagnostopoulos[1], and T. E. Matikas[1]

[1] Department of Materials Science and Engineering, University of Ioannina, Ioannina, Greece
g.varfi@uoi.gr

[2] Department of Conservation of Antiquities and Works of Art, University of West Attika, Aegaleo, Greece

[3] Institute of Nanoscience and Nanotechnology, NCSR "Demokritos", Aghia Paraskevi, Greece

Abstract. This work deals with the evaluation of post-Byzantine icon using complementary nondestructive techniques. Sampling from works of art should be avoided in most cases, therefore, it is of great importance to study the item under investigation using reliable nondestructive tools. In this study, two nondestructive evaluation techniques were used; X-Ray Fluorescence spectrometry (XRF), for performing a qualitative analysis of icon's chemical elements, and Infrared Thermography (IRT), for examining the icon's background and providing stratigraphic analysis by monitoring the thermal diffusion process in the sample following a thermal perturbation. Both techniques were able to provide useful information about the preliminary drawing under the painted surface, as well as to reveal a possible second overdrawing surface or the painter's palette. These results can provide the conservator with significant knowledge for the preservation of the artwork.

Keywords: Nondestructive evaluation · Cultural heritage · Preservation

1 Introduction

The conservation of artwork is an important aspect of preserving human tangible culture and artistic expression of different periods of time or school of artists. That goes without saying that is a multidisciplinary field and is crucial to coordinate both knowledge and technology. Works of art naturally deteriorate through years, mainly due to photooxidation processes, atmospheric pollution, possible thermal degradation or any other manual intervention (mechanical stresses etc.). Another major factor that causes several problems to the icons (i.e. religious portable panel paintings) is humidity, which is one of the main reasons for the deterioration and degradation [1]. It causes structure changes in the painting materials and also promotes the growth of micro-organisms [2].

A. Osman and A. Moropoulou (Eds.): *Advanced Nondestructive and Structural Techniques for Diagnosis, Redesign and Health Monitoring for the Preservation of Cultural Heritage*, SPM 16, pp. 16–27, 2022.
https://doi.org/10.1007/978-3-031-03795-5_2

Art conservators strive to keep artworks in good condition by performing non- invasive preservation and restoration procedures. More recently there has been an introduction of newer technologies allowing conservators to gather even more details about the artwork, permitting better knowledge of artist's materials and techniques, and as a subsequent, this will lead to a better manipulation of conservation treatments. In recent years, the systematic study of cultural heritage has established the conservation science, as one of highly interdisciplinary nature. A critical and selective eye is absolutely needed, to allow researchers to choose the most effective combination of techniques for each project and, more importantly, to process and analyze the resulting volume of diverse data [3]. Furthermore, non-contact and nondestructive techniques, nowadays, are available for diagnosing and analyzing icons or other works of art and definitely this is much appreciated by art conservators, mainly due to the fact that there is no need for sample extraction, so they are contactless and totally safe for the artwork [4].

Optical techniques are preferred in antiquities due to their nondestructive nature while on the other hand chemical analysis is tried to be avoided. The latter not only needs extraction of a sample which is destructive for a part of the artwork, even negligible, but also it may take longer time, and also needs modification and/or preparation of the samples. The former, including Infrared Thermography (IRT) and X-ray fluorescence spectroscopy (XRF), which have been employed and in the current study, are nondestructive, contactless and implementable in situ. They are capable to assess an artwork and record several measurements, in minimum time. Another major advantage is that due to the portability of the equipment, they can easily be applied in the area where the artwork is kept and in this case any risk of further damage due to transportation is avoided. These unique characteristics enable, nondestructive evaluation, as an important and useful tool for preservation and conservation of cultural heritage [5, 6].

Infrared Thermography is a real-time low cost and full field technique based on monitoring the temperature variation on the surface of materials and structures by mean of infrared radiation detection [7]. In general this method can evaluate the damage distribution and provide helpful information for restoration and conservation processes, which consequently will prevent irreversible damage of an icon or any other artwork [8, 9].

The efficiency of Infrared Thermography as a nondestructive testing and evaluation (NDT & E) technique in the literature is well documented, also in the investigation of historic structures. It has outstanding advantages in a wide span of applications, either in the assessment of structural materials and techniques (state of preservation- assessment of moisture impact), or for the evaluation of conservation interventions (materials and techniques- compatibility of restoration) [10, 11]. Furthermore, in order to obtain useful information from the infrared thermographic surveys, various properties (thermal, optical, physical) of the examined materials should be taken into account [12].

Applications of Infrared Thermography are divided in active and passive, depending on the existence or not of thermal stimulus [13, 14]. Active thermography, which requires external thermal stimulation is further categorized to: Pulsed Thermography (PT), Pulsed Phase Thermography (PPT), Lock-in Thermography (LT), etc. depending on the type of the heating type (e.g. sinusoidal, square or other type of excitation) [15–17]. In the current study Lock-in Thermography is applied. Lock-in compared to other thermal imaging approaches, provides improved thermographs due to the averaging nature of the technique and its emissivity independency [16]. The results of lock-in thermography (thermographs), have been studied for the evaluation of icon's subsurface or a possible underpainting layer and to assess the state of damage of the icon.

Another technique that has been used in the current study is X-ray Fluorescence Spectroscopy (XRF) and more precisely portable XRF (pXRF). Various instrumental and experimental modifications of XRF spectrometers meet nearly all the requirements for safe and efficient analysis of cultural heritage objects, allowing this technique to be the most used analytical method for that purpose. Nowadays, pXRF spectrometry is probably the most used portable analytical technique in the framework of heritage assets' investigation, since it enables in-situ nondestructive analysis and produces large quantity of data in a relatively short measurement time [18]. Analysis of multi-layered heterogeneous objects, like paintings and icons, is very demanding and highly complex, mostly since it is very difficult to define a reliable informative volume of the analyzed objects. As a result, only qualitative or semi-quantitative results can be derived (i.e. detection and confirmation of constituent chemical elements expressed as peak intensities) [19].

The analysis by XRF is mainly a qualitative and also quantitative method, based on the measurement of the characteristic X-ray intensities emitted by the elements that constitute a sample. X-rays emitted by X-ray tubes, which excite the elements present in the sample, so they emit spectral lines in characteristic energies. Although XRF is an excellent technique for painting analysis and it is widely used in literature, it does have some limitations. For example, because it is an elementary technique, pigments must be identified indirectly by means of key-elements associated with the colors observed in the painting [20]. However, for investigating the pigments of the painting colors of the works of art, a nondestructive analyzing method can be easily conducted by a portable XRF, aiming to trace inorganic pigments from which it consists [21].

In this study, handheld XRF had been used for the investigation purposes of the post-byzantine icons. Elemental analysis took place, in order to determine the elements from which each color of the icon is made of. Colors and areas that have been examined were selected by optical observation, depending on the different colors or hues presenting on the artwork. Most of inorganic pigments that have been used can be identified, mixtures of them or a possible overpainting layer.

Moreover, the results obtained with these two different methods, are complementary and their crosschecking brings thorough information. The collected information can be separated in two domains: imaging (Infrared Thermography) and identification (X-ray Fluorescence Spectroscopy). Both results, led to a better comprehension of painter's palette and manufacturing methods, regarding to aid conservation treatments.

2 Experimental Study

2.1 Description of the Post-Byzantine icon and its Conservation Treatments

In the current study, the post-Byzantine icon that had been studied is dated back to the second half of the 19th century, according to the inscription on the base of the artwork. It originates from Asia Minor and more precisely from Constantinople. It is painted on wooden panel and the painting surface of the icon that has been investigated, consists of three zones. In the upper zone is depicted Virgin Mary on the left side, Jesus Christ in the middle and Saint John on the right side. In the central zone are depicted four saints. First on the left is Saint Pantaleon, second is Saint Tryphon, third is Saint Catherine and the fourth figure is an unknown Saint. In the lower zone, on the left is depicted Saint George, Saint Konstantinos and Saint Helen in the middle and Saint Demetrius on the right side.

Regarding the state of preservation of the artwork, it was obvious the darkening and the craquelure of the surface due to the physicochemical degradation of the old varnish. That happens due to ultraviolet (UV) radiation of the sunlight and surface dirt. Moreover, thermal degradation signs were documented on the surface of the icon, mostly on the lower zone. Also, there was loss of painting surface on the frame of the icon on the right and lower side.

Depending on the icon's pathology, the conservation treatments that took place were, firstly the removal of the old degraded varnish, the consolidation of the frame, color retouching and finally the placement of new protective coating varnish.

For conservation treatments variant chemicals were used. For the cleaning stage Acetone (Chemco, 98,5%) was used as a solvent, for the consolidation stage acrylic emulsion Primal SF 016 (10% water solution) was used, for the color retouching pigments with a binder medium were used and for the final stage of the new coating, varnish (Talens) was also used.

2.2 Experimental Set Up

Infrared Thermography (IRT). The experimental set up consists of an IR sensor, a halogen lamp, a temperature control camera and the under investigation icon. The IR sensor operating in the infrared longwave band (LWIR) with thermal sensitivity of less than 50 mK and resolution 640 × 480 pixels. For continuous and uniform heat flux an appropriate thermal excitation source was used (halogen lamp power of 300 W). More specifically, the IR sensor was placed at the distance of 40 cm from the icon, as seen in Fig. 1. The Lock-in frequency ranged from 0.01 to 0.1 Hz and the halogen lamp amplitude was set between 40 to 80%. It should be also mentioned that in order to eliminate any possibility of heating the icon above a critical temperature (in our case 30 °C according to the instructions of the scientific group conservator), a second mid-wave IR camera was continuously recording the specimen temperature.

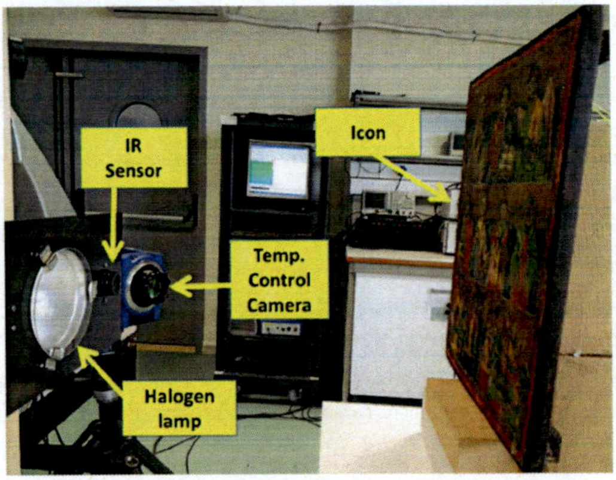

Fig. 1. Experimental set up of IR thermography

X-ray Fluorescent Spectroscopy (XRF). A portable handheld XRF spectrometer (Tracer 5i, Bruker), equipped with Rh target X-ray tube and a silicon drift detector, was used to conduct elemental analysis. The energy resolution of the detector was typically 140 eV at the Mn Kα-energy. The measurements were carried out for a total of 30 s and the working conditions were 15 μA at 50 kV and a spot size of 3 mm (Fig. 2).

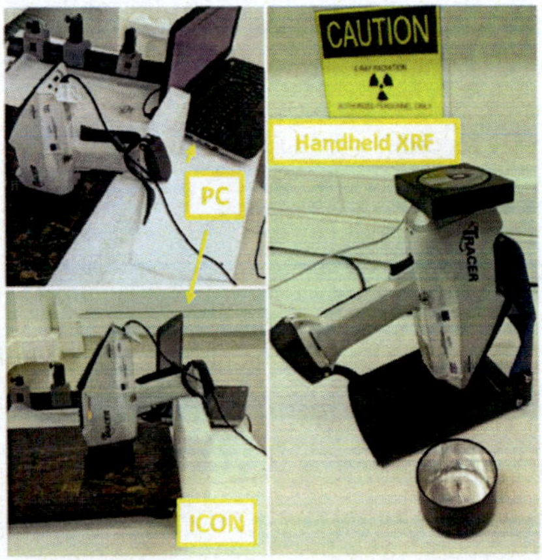

Fig. 2. Experimental set up of X-ray fluorescent spectroscopy.

3 Results and Discussion

Lock-in thermography was applied in several areas of interest in the under-investigation icon. The first area of interest is shown in Fig. 3a, where Virgin Mary is represented. According to the IR thermograph (at 0,1 Hz, 80% amplitude) Fig. 3b and the magnification Fig. 3c it can be clearly observed the existence of Greek letter "P" (equivalent to R in English alphabet) which is not clearly visible with naked eye and in addition with letter "M" that is visible even in the digital photo, validates that the representation is "the Mother of the God"(MP = MHTHP).

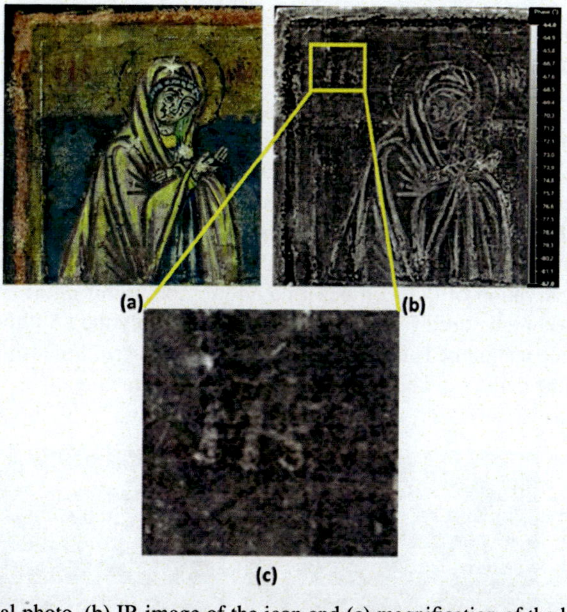

Fig. 3. (a) Digital photo, (b) IR image of the icon and (c) magnification of the highlighted area where can be observed the detail of the Greek letters "MP" meaning mother

Remarkable is the Fig. 4a which represents Jesus. In the thermograph in Fig. 4b (at 0,01 Hz, ~40% amplitude) can be observed humidity at the point of the head in Jesus. Also, the thermograph in Fig. 4c with different lock-in parameters (0.1 Hz frequency and 80% amplitude), reveals intense humidity. Specifically the area inside the yellow circle is the only area where the average phase is 64,6°, while other 'white' spots like the gospel letters the average phase is about 68,9°. These findings inform early and without even touching the icon, the conservator that this specific point needs special treatment. Moreover, in Fig. 4a the first letter at the gospel that Jesus holds in his hand, is not visible, while in the thermograph Fig. 4c of the same area it is obvious that the first letter is Epsilon "E".

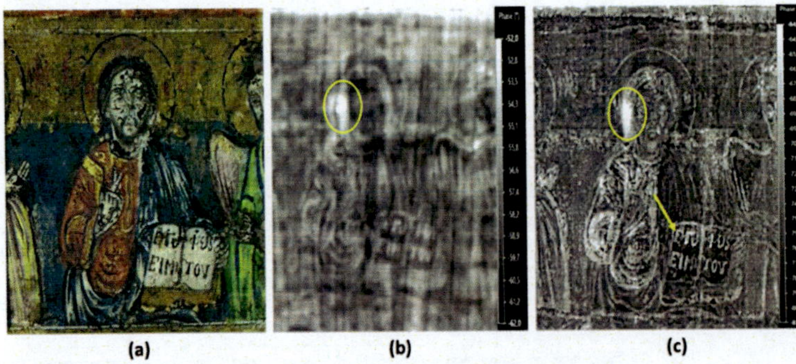

Fig. 4. (a) Digital photo, (b) IR image at 0.01 Hz and (c) IR image at 0.1 Hz of the icon where Jesus is presented.

Another area of the icon that has been studied which represents an unidentified Female Saint (could be Saint Paraskevi) is shown in Fig. 5a. In the digital photo of the Saint can be observed only the excessive surface damage, but at the IR thermogram (at 0,1 Hz, 80% amplitude) Fig. 5b the preliminary drawing is clearly revealed below the painting (e.g. yellow arrows), and therefore can help the conservator not only in the conservation of the icon but it could provide useful information about the history or the technique used for the specific icon (probably the "Anthivolo" Greek Orthodox Byzantine technique is used). This is further supported, because in the right side of the mantle the painter did not follow the preliminary drawing.

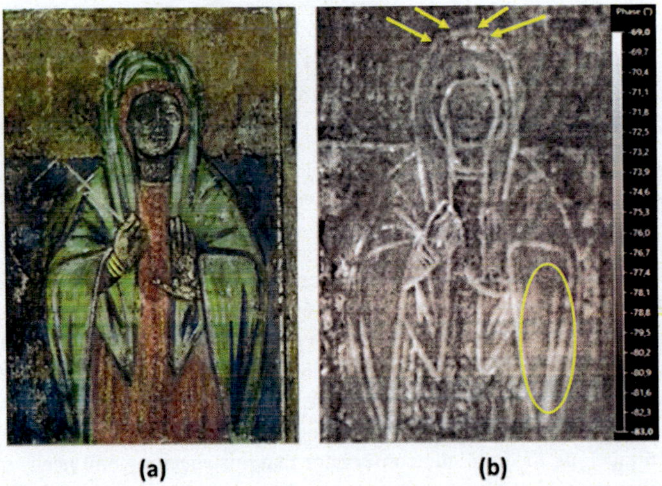

Fig. 5. (a) Digital photo and (b) IR image of the icon at 0.1 Hz where one unknown saint is presented.

The last area that was examined with IR thermography was the Saint Demetrius (Fig. 6a). According to the IR image at 0.1 Hz and 80% amplitude Fig. 6b again the preliminary drawing of the military uniform and especially the contouring of the cape can be observed. This is significant due to the fact that the cape is in bad shape as can be observed at the digital image Fig. 6a and the contours will provide the required information in order to restore the icon as closely to the original as possible. It should also be mentioned that the preliminary drawing of the horse saddle is clearly depicted in Fig. 6b while the painter did not followed it exactly, as seen in Fig. 6a.

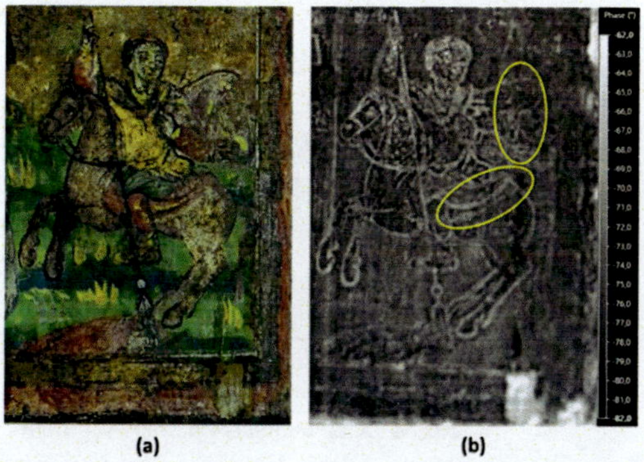

Fig. 6. (a) Digital photo and (b) IR image of the icon at 0.1 Hz which is presented the saint Demetrius

Furthermore, X-Ray Fluorescence (XRF) elemental analysis was performed to the pigments used in the areas of interest that were examined above with IR Thermography. Firstly, where Virgin Mary is depicted, the measurement was performed on the red letters, which are on the left side of Virgin Mary's head (MP). According to the spectrum, the detection of mercury (Hg) and sulfur (S) indicates the employment of cinnabar (HgS) in the red letters. The simultaneous detection of intense lead (Pb) peaks may pertain to cinnabar adulteration with red lead/minium, a practice recently documented in relevant artworks (Fig. 7).

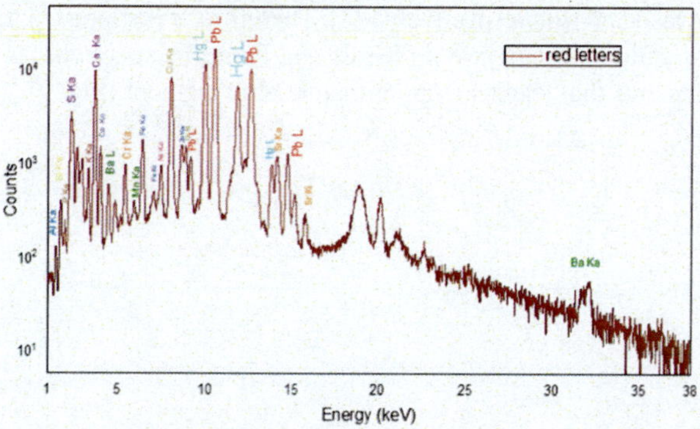

Fig. 7. XRF analysis on red pigment.

In the case of Jesus, which was the second area of interest that was examined, the measurement was performed to the black letters of the gospel, which Jesus holds on his left hand. We notice at the spectrum that lead (Pb) gives a very strong peak, which comes from the white background behind the letters of the gospel (white lead-$2PbCO_3{\cdot}Pb(OH)_2$). As elements characteristic to known black pigments (e.g. phosphorous, manganese) are not detected in the letters, one may assume that the black letters have been rendered of an organic black pigment (Fig. 8).

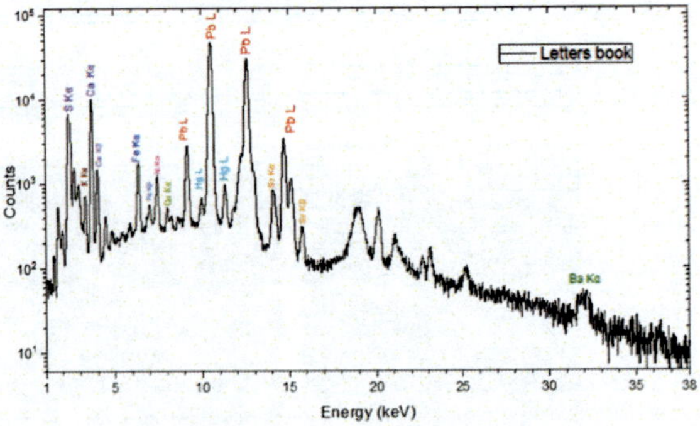

Fig. 8 XRF analysis on black pigment

Moreover, in the third case of the female unknown saint, the pigments that were used for the mantle, were also measured with XRF. Hereby, it is important to notice that these pigments seemed to have more barium (Ba) than any other pigment of the icon. Blue and green are the two pigments, which were measured, on the left's hand mantle of the unknown saint. Many different spots and colors of the icon were measured and according to the spectrums the elements seem to be similar, with minor differences in each spot or color. But, in these two colors (blue-green), barium was definitely in greater quantity than in the other pigments. According to the spectrums, we suppose that the absence of copper in blue color may either indicate employment of an organic-type pigment (e.g. indigo) or Prussian blue (detection of minor Fe) [22]. Green color probably comes from a mixture of blue and yellow. Yellow color according to the measurements, seem to come from chrome yellow ($PbCrO_4$) and yellow ochre ($FeO(OH)$), based on the detection of chrome, lead and iron respectively [23]. Therefore, it is supposed that green comes from blue and yellow because we notice Cr and Fe at the spectrum, as well as Ba, and any other elements that lead us to an inorganic blue pigment (Fig. 9).

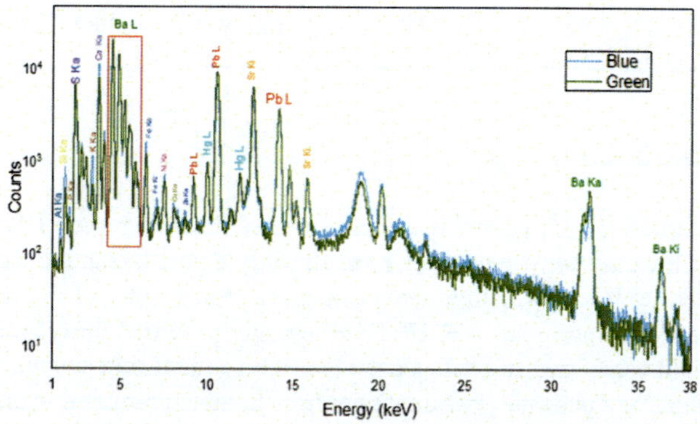

Fig. 9. XRF analysis on blue and green pigments.

The last areas that were examined, was on Saint-Demetrius cape and horse's saddle. For the rendering of the light purple-pink color of the cape, it appears that a pigment of organic origin (i.e. lake) was used, That light purple-pink color of the cape, also according to the measurements seem to be organic pigment, as can't be noticed any inorganic element in greater amount, as, apart from lead (which probably stems from Pb-white admixtures). As for the red and green colors of horse's saddle, are the same pigments as these ones that had been used above. Red is cinnabar (Virgin Mary) and green is the same (mixture blue - yellow) as the one used in the unknown saint's mantle (Fig. 10).

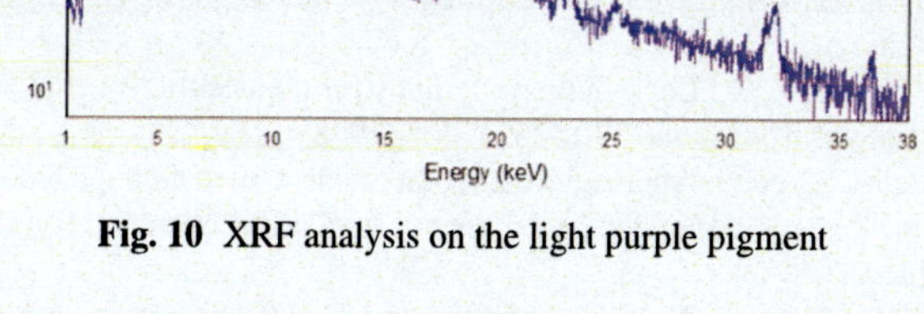

Fig. 10 XRF analysis on the light purple pigment

Combining the results obtained using IR Thermography and XRF, it enabled to reveal the underdrawing that the artist applied for the painting of the icon, as well as the pigments that were used. More specifically, the thermographs indicated, in addition to the absorbed humidity present same spots which need to be addressed by the conservator, also information about the painter's underdrawing which are well-depicted in the thermograph even if the painting's surface has been destroyed. Furthermore, on the same areas of interest, where IR Thermography was applied, XRF elemental analysis evaluated the pigments that were used for painting the icon.

4 Conclusion

Nondestructive techniques, including Infrared Thermography (IRT) and X-ray Fluorescent Spectroscopy, have been used to study a post-byzantine icon. Both techniques provided useful information about either the stratigraphy of the icon or the pigments used in the painting surface. IR thermography provided useful information and hints that could verify and authenticate the icon's manufacturing technique (or even the manufacturer), or a possible post-treatment from another conservator, enabling to decide the optimum conservation treatment for the specific artifact. Additionally, elemental analysis provided significant information about the pigments used in the icon. The painter's palette could be studied to a satisfactory point, helping to assess about the specific inorganic pigments the artist used to create the artwork, as well as to make assumptions of the possible presence of the organic ones.

References

1. Issa, Y.M., et al.: A combination of analytical methods to evaluate the effect of humidity aging on the painting materials of icon models. Vib. Spectro. **107**, 103010 (2020)
2. Liritzis, I., et al.: Archaeometry: an overview. Sci. Cult. **6**(1), 49–99 (2020)
3. Stratis, J.A., et al.: Enhancing the examination workflow for Byzantine icons: implementation of information technology tools in a traditional context. J. Cult. Herit. **15**(1), 85–91 (2014)
4. Borg, B., et al.: The application of state-of-the-art technologies to support artwork conservation: literature review. J. Cult. Herit. **44**, 239–259 (2020)
5. Elias, M., Mas, N., Cotte, P.: Review of several optical nondestructive analyses of an easel painting. Complementarity and crosschecking of the results. J. Cult. Heritage **12**(4), 335–345 (2011). https://doi.org/10.1016/j.culher.2011.05.006
6. Brunetti, B.G., Matteini, M., Miliani, C., Pezzati, L., Pinna, D.: MOLAB, a mobile laboratory for in situ non-invasive studies in arts and archaeology. In: Nimmrichter, J., Kautek, W., Schreiner, M. (eds.) Lasers in the Conservation of Artworks. Springer proceedings in physics, vol. 116, pp. 453–460. Springer, Heidelberg (2007). https://doi.org/10.1007/978-3-540-72310-7_54
7. Wu, D., Busse, G.: Lock-in thermography for nondestructive evaluation of materials. Revue Générale de Thermique **37**(8), 693–703 (1998)
8. Kordatos, E., et al., Application of IR thermography to damage characterization of structures and the diagnosis of historic monuments. In: Construction and Building Materials, pp. 77–81 (2012)
9. Kordatos, E.Z., et al.: Infrared thermographic inspection of murals and characterization of degradation in historic monuments. Constr. Build. Mater. **48**, 1261–1265 (2013)
10. Moropoulou, A., et al.: Infrared thermographic assessment of materials & techniques for the protection of cultural heritage. Proc SPIE **4548**, 313–318 (2001)
11. Ibarra-Castanedo, C., Maldague, X.P.V.: Infrared thermography. In: Czichos, H. (ed.) Handbook of Technical Diagnostics: Fundamentals and Application to Structures and Systems. Springer, Heidelberg, pp. 175–220 (2013). https://doi.org/10.1007/978-3-642-25850-3

12. Avdelidis, N.P., Moropoulou, A.: Applications of infrared thermography for the investigation of historic structures. J. Cult. Herit. **5**(1), 119–127 (2004)
13. Yao, Y., et al.: Active thermography testing and data analysis for the state of conservation of panel paintings. Int. J. Therm. Sci. **126**, 143–151 (2018)
14. Farmaki, S., et al.: A novel infrared thermography sensing approach for rapid, quantitative assessment of damage in aircraft composites. Sensors **20**(15), 4113 (2020)
15. Rocha, J.H.A., Santos, C.F., Póvoas, Y.V.: Evaluation of the infrared thermography technique for capillarity moisture detection in buildings. Procedia Struct. Integrity **11**, 107–113 (2018)
16. Rasha, L., et al.: Water distribution mapping in polymer electrolyte fuel cells using lock-in thermography. J. Power Sources **440**, 227160 (2019)
17. Moropoulou, A., et al.: Nondestructive techniques as a tool for the protection of built cultural heritage. Constr. Build. Mater. **48**, 1222–1239 (2013)
18. Andrić, V., et al.: Evaluation of pattern recognition techniques for the attribution of cultural heritage objects based on the qualitative XRF data. Microchemical J. **167**, 106267 (2021)
19. Bonizzoni, L., ED-XRF analysis for cultural heritage: is quantitative evaluation always essential? J. Phys. Conf. Ser. **630**, 012001 (2015)
20. Molari, R., C.R. Appoloni, Pigment analysis in four paintings by Vincent van Gogh by portable x-ray fluorescence (pXRF). Radiat. Phys. Chem. **181**, 109336 (2021)
21. Ha, J.-W., Lee, S.-J.: Identification of natural inorganic pigments used on 18th century Korean traditional mural paintings by using a portable X-ray fluorescence. J. Ind. Eng. Chem. **28**, 328–333 (2015)
22. Mastrotheodoros, G.P., Beltsios, K.G., Bassiakos, Y.: On the blue and green pigments of post-byzantine Greek icons. Archaeometry **62**(4), 774–795 (2020)
23. Mastrotheodoros, G.P., Beltsios, K.G., Bassiakos, Y.: On the red and yellow pigments of post-byzantine Greek icons*. Archaeometry **63**(4), 753–778 (2021)

Reflectance Spectroscopy as a Non-invasive Technique for Preliminary Characterization of Soluble Salts on Historical Masonries

Dafne Cimino(✉), Laura Falchi(✉), Martina Corradini, Paola Lucero Gomez, Eleonora Balliana, and Elisabetta Zendri

Ca' Foscari University of Venice, via Torino 155, 30172 Venice, Italy
{dafne.cimino,laura.falchi}@unive.it

Abstract. The presence of soluble salts due to rising damp triggers significant damage to brick masonries, and their distribution needs to be effectively assessed before targeting maintenance intervention. For this purpose, a novel non-invasive approach based on reflectance spectroscopy analysis has been proposed for the preliminary characterisation of soluble salts on masonry surfaces by using a portable spectroradiometer, working in the visible-short wave infrared (Vis-SWIR, 350 ÷ 2500 nm) spectral range. The setup of the technique included the analysis of powdery samples: pure salts, salts mixtures, salts blended with brick powder. Then on field application on a Venetian palace, subjected to rising damp of salty water, was carried out. The results highlight the feasibility of the methodology both in the laboratory and *in situ*: thanks to instrument portability, low cost, and an adequate possibility to detect salts this technique helps in the planning of the preservation project.

Keywords: Reflectance spectroscopy · Non-invasive method · Soluble salt

1 Introduction

The transport, deposition, and accumulation of soluble salts within porous building materials led to accelerated degradation of masonries causing powdering, crumbling and detachments of bricks and mortars [1–4]. Before any intervention, a proper evaluation of the conservation state of the building, also concerning the presence and distribution of soluble salts, is crucial. In fact, detecting salts spatial distribution and composition in the masonries can avoid a precocious failure of the intervention, such as plaster detachment or decohesion [5]. Commonly, the presence of salts is evaluated and mapped by direct visual observation, however environmental temperature and relative humidity conditions might affect their detection. Thermal imaging could help, indirectly, in the identification of their presence pinpointing humid areas along masonries. This technique is, however, affected by the surface temperature of the facade and data interpretation needs to be aware of this sensitivity. Moreover, both methods, *in situ* visual observations and thermal analyses, do not provide specific information on salts composition [6, 7], which requires

A. Osman and A. Moropoulou (Eds.): *Advanced Nondestructive and Structural Techniques for Diagnosis, Redesign and Health Monitoring for the Preservation of Cultural Heritage*, SPM 16, pp. 28–37, 2022.
https://doi.org/10.1007/978-3-031-03795-5_4

invasive punctual sampling and a full laboratory analysis boosting times and costs [1, 5, 8, 9].

An interesting alternative might be represented by non-invasive and *in situ* reflectance spectroscopy, at least for a first screening on the facade.

This technique is largely employed for soil characterisation [10–13] and evaluation of the humidity and salt content in several matrices such as ground or sand [10, 14–16]. Its applications in the conservation field are so far limited to pigments identification [17], evaluation of degradation processes on carbonate-based stones [18, 19] or attempts of monitoring the conservation state of historical buildings facades [20].

This research aims at understanding and verifying the possible application of reflectance spectroscopy (RS) as a useful technique for evaluating the presence and distribution of soluble salts in mansories. The detection and characterization of soluble salts over building materials is, obviously, a photography of the conservation state of the building itself at a specific moment. Nevertheless, RS, carried out in the IR range, can be a reasonable and useful resource for understanding salts origin (i.e. exogen or endogen) and distribution. Their chemical identification and distribution can help, moreover, in evaluating the evolution of the degradation processes.

In the laboratory, different salts commonly related to rising damp processes of sea-water were analysed on their own, at different hydration states, and in binary mixtures. Later, the powder of modern bricks, generally used in the repair of traditional masonry, was added to a quaternary-salt blend (Na_2CO_3, Na_2SO_4, KNO_3 besides NaCl) to discriminate the contribution of this common building material to the reflectance curve. In parallel, the analysis of powdery real samples, collected by drilling of brick masonries, was carried out.

The laboratory results encouraged to test this technique on masonry subjected to rising damp and loaded with soluble salts in Ca' Foscari Palace, a Venetian historical building already studied in the past [21].

Analyses were carried out with a portable spectroradiometer working in the visible-short wave infrared spectral range ($350 \div 2500$ nm). Reflectance spectra collected in this broad range provide double information: physical, related to the colour of the material and chemical due to characteristic absorptions of the present molecules, thus allowing the identification of several organic and inorganic compounds [17, 18, 20, 22–28].

2 Materials and Methods

2.1 Analysis of Pure Soluble Salts and their Mixtures

Pure soluble salts, some of them commonly present as efflorescence on masonries (Na_2CO_3, Na_2SO_4, KNO_3, $CaSO_4{\cdot}2H_2O$, Na_2SO_4, NaCl) and others associated with building materials ($CaCO_3$, $MgCO_3$, $CaMg(CO_3)_2$) were analysed by reflectance spectroscopy on their own. $CaCl_2$, $BaCl_2{\cdot}2H_2O$ were also considered as commonly used in winter in urban areas thanks to their defrosting capabilities.

Considering that sodium chloride (NaCl) is the most common salt found in masonries in coastal areas, binary mixtures with it, as major components, were tested as well to evaluate the detection limit of the instrument. NaCl amount ranged between 80 and 97.5% w/w. The second component was $CaCl_2$, KNO_3, Na_2CO_3 respectively.

A quaternary mixture was later considered: it was formulated based on the average concentration of soluble salts found in historical Venetian masonries in previous studies, in particular concerning the ratio between Cl^- and SO_4^{2-} [29, 30]; carbonate and nitrate were added as well in an amount comparable with the sulphate [31]. Therefore, Na_2CO_3: Na_2SO_4: KNO_3: NaCl were mixed in ratio 1:1:1:11; the powder of a modern brick (from 5 up to 95% w/w, 5% steps) was then added to the quaternary mixture. The red modern bricks were supplied by San Marco Terreal s.r.l. as these are often used in restoration interventions in Venice.

All the pure salts were supplied by VWR Chemicals.

For obtaining reliable spectra of the hydrated species, $Na_2CO_3{\cdot}7H_2O$, $MgSO_4{\cdot}7H_2O$, $Na_2SO_4{\cdot}7H_2O$ standards were kept in the laboratory (ca 25 °C) for one month at different relative humidity conditions (~30, 69, 81%) before exploring their reflectivity according to their hydration level (Table 1).

Table 1. Hydrated states of the considered soluble salts

Salt	Phases
$MgSO_4$ [32]	Kieserite (1 H_2O) up to ca 40% RH Hexahydrite (6 H_2O) ca 40% < RH < 50% Epsomite (7 H_2O) RH > ca 50%
Na_2CO_3 [33]	Higher hydration level above 60% RH
Na_2SO_4 [34]	Thenardite up to 75% RH Mirabilite (10 H_2O) RH > 75%

2.2 Real Cases

Bricks of a pillar in the main atrium of Ca' Foscari Palace (Venice, Italy) were analysed *in situ* by reflectance spectroscopy (RS). Masonries of this building are particularly rich in soluble salts either as efflorescence or sub-efflorescence. Four bricks (at 20, 86, 140 and 197 cm from the walking level respectively) were also sampled by drilling considering four depth sampling steps (0 ÷ 2, 2 ÷ 5, 5 ÷ 10, 10 ÷ 15 cm); powders were lately analysed with RS in the laboratory.

2.3 Spectroradiometric Analysis

Analyses were carried out with a portable ASD Fieldspec 4 Hi-res (Malvern Panaliytical) spectroradiometer, working in the visible-short wave infrared (Vis-SWIR) spectral range, from 350 to 2500 nm (resolution: 3 nm between 350 ÷ 1000 nm, 8 nm between 1000 ÷ 1800 and 1800 ÷ 2500 nm), and each spectrum was the average of 100 measurements. A two-step calibration of the instrument preceded the analytical session: acquisition of the instrumental black current was the average of 25 measurements followed by the optimization of the three detectors and the acquisition of the white signal, analysing a certified Spectralon®. The choice of a contact probe with an inner halogen light source guarantees a more homogeneous lightening of the investigated surfaces and avoid possible interferences due mainly to environmental light. This probe ensures, also for powder, a higher reproducibility as the analysis is less affected by sample inhomogeneities.

3 Results and Discussion

3.1 Analysis of Pure Soluble Salts and their Mixtures

As in other spectroscopic techniques (such as Infrared and Raman spectroscopies) [35], NaCl is not recognizable apart from the retained water due to its hygroscopic nature. Similarly, potassium and magnesium sulphates (K_2SO_4, $MgSO_4{\cdot}7H_2O$) do not exhibit a characteristic spectrum. Clearly detectable are instead barium and calcium chlorides ($CaCl_2$, $BaCl_2{\cdot}2H_2O$), calcium, magnesium, calcium-magnesium and sodium carbonates ($CaCO_3$, $MgCO_3$, $CaMg(CO_3)_2$, Na_2CO_3), calcium and sodium sulphates in their hydrated states ($CaSO_4{\cdot}2H_2O$, $Na_2SO_4{\cdot}10H_2O$) and magnesium and potassium nitrates ($Mg(NO_3)_2$, KNO_3). In terms of reflectivity, salts with higher water content give spectra with a lower total intensity. Magnesium sulphate did not show changes in its spectral features at different hydration states, whereas the other two compounds tested (sodium carbonate and sodium sulphate) developed absorptions that allow the possibility to discriminate their hydration states (Fig. 1).

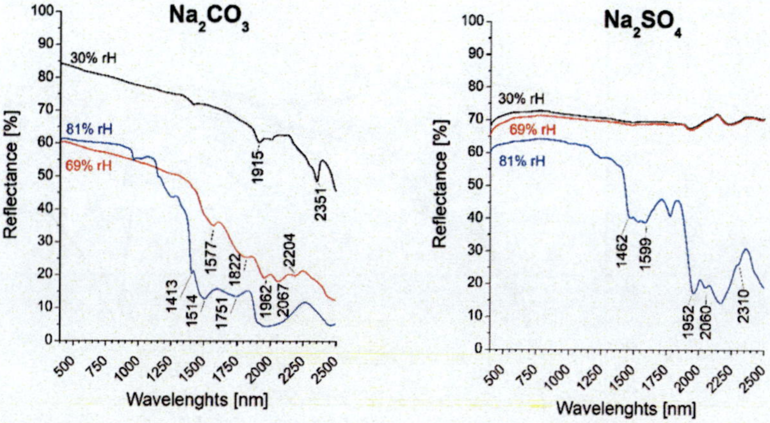

Fig. 1. Reflectance spectra of the different hydration states of sodium carbonate (left) and sodium sulphate (right)

Once pure soluble salts were pinpointed, binary mixtures of these with NaCl as the major component were analysed. In general, the presence of a second component had effects on the NaCl curve, nevertheless, the identification of the added salt was not always possible, like for the blends with Na_2CO_3 (Fig. 2, left). On the contrary, the lowest amount of KNO_3 was detectable and the minimum at 2464 nm is the identification feature (Fig. 2, right).

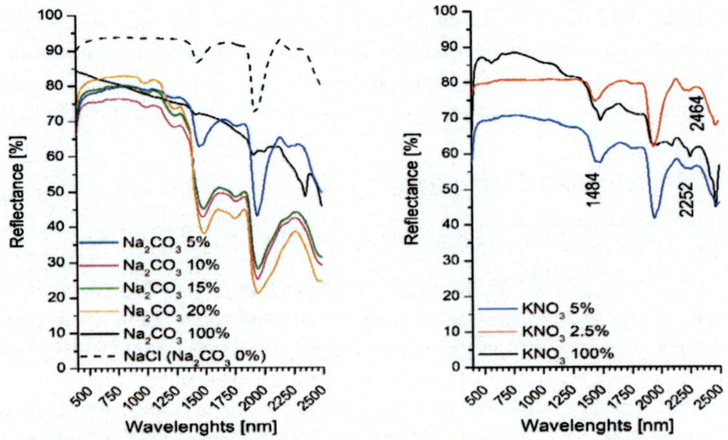

Fig. 2. Binary mixtures made of NaCl as major components with sodium carbonate (left) or potassium nitrate (right) at different concentrations.

Powder of the modern brick was added (from 5 up to 95% w/w, 5% steps) to a quaternary salt blend (QB) made of $NaCO_2$: $NaSO_4$: KNO_3: NaCl in ratio 1:1:1:11 (Fig. 3).

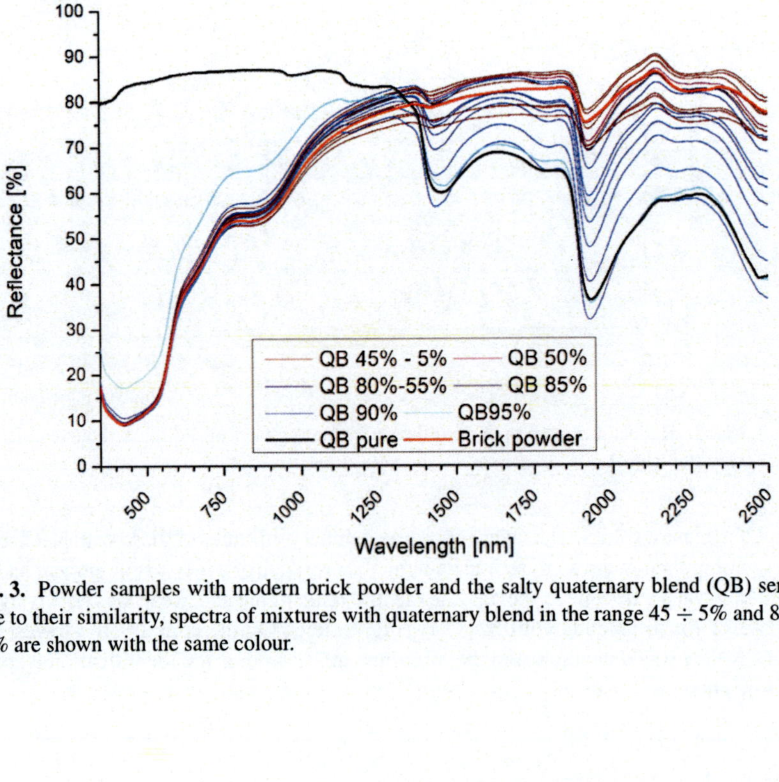

Fig. 3. Powder samples with modern brick powder and the salty quaternary blend (QB) series. Due to their similarity, spectra of mixtures with quaternary blend in the range 45 ÷ 5% and 80 ÷ 55% are shown with the same colour.

Pure compounds (QB100% and brick100%) are reported in black and red respectively. Signals due to water absorption at ca 1450 and 1950 nm covered most of the spectral features of the salts, even though the contribution of KNO_3 was recognized in the spectrum minima at 960, 1020, 2470 nm and a shoulder at 2250 nm (Fig. 3, black line). The shoulder at 2470 nm, representative of KNO_3, was detectable up to 90% w/w of the quaternary salty mixtures, i.d. 6.3% on the total weight. As so, the limit of detection for this compound, fixed previously at 2.5% in the binary mixture, was here much higher.

As pointed out by Fig. 3 even a minimum amount of the brick powder caused a clear shift to a red hue of the mixture, as observable in the Vis range: the relative spectrum (Fig. 3, light blue line) was very similar to the brick one (Fig. 3, red line) with characteristic features at 520, 650, 750 and 880 nm due to the presence of iron oxides, in particular hematite [12].

The colour of the brick-salts system resulted soon saturated, and the relative spectra are not distinguishable in the range 350 ÷ 750 nm as they overlapped; in the SWIR region, instead, the influence of the content of humidity decreased and the associated signals shifted to lower wavenumbers because associated with the structural water of the clay minerals.

3.2 Analysis of Real Cases

An attempt of identification of salts was carried out on real samples from a pillar of the historical palace of Ca' Foscari in Venice.

In situ validation of the methodologies, ensuring a greater contact with the efflorescence allowed the identification of the main species present in the salt patina. This was the case (Fig. 4a-b) of the lower area interested by efflorescences (ca 50 cm from the floor), where gypsum and calcium-magnesium carbonate were identified, and of the upper level (ca 180 cm high) where along with nitrates and carbonates (2475 nm for NO_3^-, 2254 and 2300 for CO_3^{2-}), other compounds not recognised (signals at 1678, 1714, 2132 nm) were detected.

The RS analysis of powders, sampled by drilling from four bricks at 20, 86, 140 and 197 cm, did not allow the identification of the salts, neither the detection of their presence. Figure 4c reports the RS spectra of samples C9 and C12 collected on the same brick at different depths (0 ÷ 2 and 5 ÷ 10 cm respectively). These samples are rich in soluble salts, as detected by conductivity measures, but the fine grinding of the powder during the sampling made practically indistinguishable the salts contribution in the spectra. On the contrary, when salt grains were still clearly visible, the RS probe was able to identify their chemical nature (Fig. 4d). For example, in sample C24, particularly reach in salts, but not homogeneously ground, the spectrum allowed a better identification, in particular of KNO_3 and $MgCO_3$.

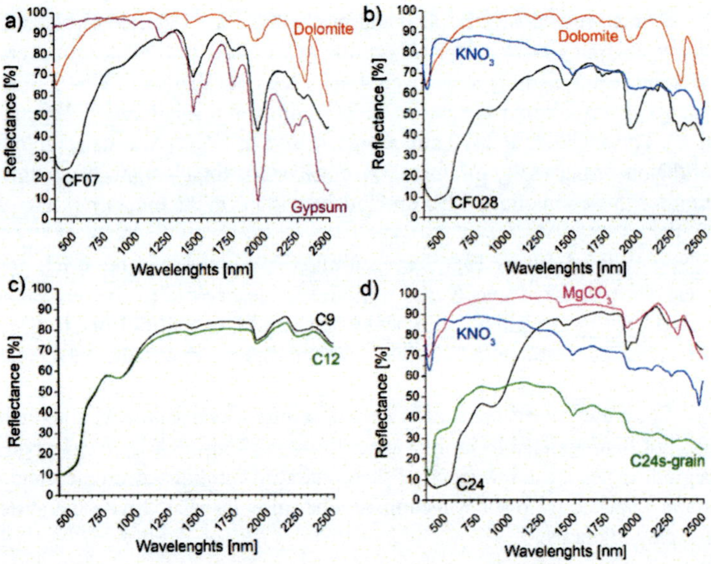

Fig. 4. Reflectance spectra of bricks of the pillar in Ca' Foscari palace. a), b) RS spectra from a lower (CF07) and a higher (CF028) areas with efflorescences along with gypsum, dolomite - Mg,Ca(CO_3)- and KNO_3 references spectra; c), d) RS spectra of powders sampled along the pillar from historical bricks of Ca' Foscari palace

4 Conclusion

Data obtained showed that sodium chloride remains a salt difficult to be recognised by spectroscopic techniques, revealing its presence in reflectance spectra only in force of its hygroscopic behaviour. Despite that, other salts among sulphates, chlorides, carbonates and nitrates are easily recognised and, in some cases, the limit of detection of the species is lower than 2.5% when present in binary mixtures associated with NaCl. The hydration state of some salts might be characterised as well. The presence of brick powder, either in the case of laboratory blends or in real cases, lowered the relative detectable salt concentration.

Nevertheless, the analyses carried out on powders sampled from real cases (masonry in Ca' Foscari) showed that it is still possible to identify some of the salts present, such as potassium nitrate and magnesium carbonate. Efflorescences along masonry analysed *in situ* were identified, when possible, thanks to a direct comparison with an *ad hoc* database, obtained by in-lab analyses of pure salts. The validation of the methodology for real applications highlighted the need for the collection of reference spectra, which with a direct comparison through the same software simplify the identification of the compounds. Indeed, spectra are not just the sum of the signals of the compounds that are in the samples, and comparison limited to literature figures is difficult, considering that each real case is a different situation.

Being a quick and low-cost analysis, the obtained information is already useful for the design of a preservation project or a conservation treatment. Further analyses can be carried out based on the data collected, optimizing the sampling procedure and detecting further conservation issues, such as the presence of humidity within the masonry and its distribution in it. If, from one side, this technique can be applied to determine the conservation state of the building materials, the same analytical campaigns might allow overtime proper monitoring, visualizing how degradative phenomena proceed and identifying new degradation issues at an early stage such as new external salts deposition as efflorescence or raise of fungal and other bio-colonies.

Funding. Scientific activity performed with the contribution of the Provveditorato for the Public Works of Veneto, Trentino Alto Adige and Friuli Venezia Giulia, provided through the concessionary of State Consorzio Venezia Nuova and coordinated by CORILA. The research was promoted thanks also to the interest and support of the Patto per lo Sviluppo della Città di Venezia (Comune di Venezia).

References

1. Lubelli, B., et al.: Experimental research on hygroscopic behaviour of porous specimens contaminated with salts. Constr. Build. Mater. **18**(5), 339–348 (2004)
2. Snethlage, R., Wendler, E.: Moisture cycles and sandstone degradation. In: Snethlage, R., Baer, N.S. (eds.) Saving our Architectural Heritage: the Conservation of Historic Stone Structures, pp. 7–24. Wiley, Chichester (1997)
3. Moropoulou, A., et al.: Preservation planning as a tool for a sustainable historic city. WIT Trans. Ecology Environ. **39**, 327–336 (2000)
4. Theoulakis, P., Moropoulou, A.: Salt crystal growth as weathering mechanism of porous stone on historic masonry. J. Porous Mater. **6**(4), 345–358 (1999)
5. Nunes, C., et al.: Study of nitrate contaminated samples from a historic building with the hygroscopic moisture content method: contribution of laboratory data to interpret results practical significance. J. Cult. Herit. **30**, 57–69 (2018)
6. Giovannacci, D., et al.: Non intrusive tools to detect salts contamination in masonry: case study of fontaine-chaalis church. In: Optics for Arts, Architecture, and Archaeology VI, vol. 10331, p. 1033103. International Society for Optics and Photonics (2017)
7. Avdelidis, N.P., Moropoulou, A.: Applications of infrared thermography for the investigation of historic structures. J. Cult. Herit. **5**(1), 119–127 (2004)
8. EN 16682:2017. Conservation of cultural heritage - Methods of measurement of moisture content, or water content, in materials constituting immovable cultural heritage. CEN, Comité européen de normalisation Norme européenne (2017)
9. Nasraoui, M., Nowik, W., Lubelli, B.: A comparative study of hygroscopic moisture content, electrical conductivity and ion chromatography for salt assessment in plasters of historical buildings. Const. Build. Mater. **23**(5), 1731–1735 (2009)
10. Cipra, J.E., et al.: Measuring radiance characteristics of soil with a field spectroradiometer. Soil Sci. Soc. Am. J. **35**(6), 1014–1017 (1971)
11. Stenberg, B., et al.: Visible and near infrared spectroscopy in soil science. In: Sparks, D.L. (ed.) Advances in Agronomy, vol. 107, pp. 163–215. Academic Press, Burlington (2010)
12. Fang, Q., et al.: Visible and near-infrared reflectance spectroscopy for investigating soil mineralogy: A review. J. Spectr. **2018**, 1–14 (2018)
13. Farifteh, J., et al.: Spectral characteristics of salt-affected soils: a laboratory experiment. Geoderma **145**(3–4), 196–206 (2008)

14. Edwards, B.L., et al.: Simple infrared techniques for measuring beach surface moisture. Earth Surf. Proc. Land. **38**(2), 192–197 (2013)
15. Ma, C., et al.: Analysis of the effects of salt content on soil spectral characteristics. In: Proceedings of the 3rd International Conference on Agro-Geoinformatics, pp. 1–6. IEEE (2014)
16. Monnard, A., et al.: Influence of moisture content on colour of granular materials. Part I: experiments on yellow ochre. Granular Matter **18**(3), 47 (2016)
17. Bacci, M., et al.: Non-invasive spectroscopic measurements on the Il ritratto della figliastra by Giovanni Fattori: identification of pigments and colourimetric analysis. J. Cult. Herit. **4**(4), 329–336 (2003)
18. Camaiti, M., et al.: Hyperspectral sensor for gypsum detection on monumental buildings. J. Geophys. Eng. **8**(3), S126–S131 (2011)
19. Suzuki, A., et al.: Laboratory study of the sulfation of carbonate stones through SWIR hyperspectral investigation. J. Cult. Herit. **32**, 30–37 (2018)
20. Benvenuti, M., et al.: A portable hyper spectral device for monitoring the chemical and mineralogical composition of historical buildings surfaces. In: Ferrari, A. (ed.) 4th International Congress on Science and Technology for the Safeguard of Cultural Heritage in the Mediterranean Basin, pp. 167–168. Il Cairo, Egypt (2009)
21. Falchi, L., et al.: Rising damp in historical buildings: a venetian perspective. Build. Environ. **131**, 117–127 (2018)
22. Alparone, L., et al.: Hyperspectral instruments as potential tools for monitoring decay processes of historical building surfaces. In: Fioravanti, M., Mecca, S. (eds.) The Safeguard of Cultural Heritage: A Challenge From the Past for the Europe of Tomorrow, pp. 192–194. Firenze University Press, Firenze (2011)
23. Bacci, M.: Fibre optics applications to works of art. Sens. Actuators B Chem. **29**(1–3), 190–196 (1995)
24. Balas, C., et al.: A novel hyper-spectral imaging apparatus for the non-destructive analysis of objects of artistic and historic value. J. Cult. Herit. **4**, 330–337 (2003)
25. Camaiti, M., et al.: Monitoring of chemical and physical characteristics of stone surfaces by a portable spectroradiometer. Geophys. Res. Abstr. **15**, 13552 (2013)
26. Campbell, J.B., Wynne, R.H.: Introduction to Remote Sensing. Guilford Press, New York (2011)
27. Clark, R.N., et al.: High spectral resolution reflectance spectroscopy of minerals. J. Geophys. Res. Solid Earth **95**(B8), 12653–12680 (1990)
28. Clark, R.N.: Spectroscopy of rocks and minerals, and principles of spectroscopy. Manual Remote Sens. **3**(3–58), 2–2 (1999)
29. Biscontin, G., et al.: Indagine sui fenomeni di risalita capillare nelle murature di Venezia. In: Il mattone di Venezia. Contributi presentati al concorso di idee su patologia, diagnosi e terapia del mattone di Venezia, pp. 33–65
30. Fassina, V., et al.: Evaluation of compatibility and durability of a hydraulic lime-based plaster applied on brick wall masonry of historical buildings affected by rising damp phenomena. J. Cult. Herit. **3**(1), 45–51 (2002)
31. Corradini M.: Investigation of the rising damp phenomenon in historical Venetian buildings by a new multianalytical approach. Ms thesis, Ca' Foscari University of Venice (2020)
32. Steiger, M., et al.: Hydration of $MgSO_4{\cdot}H_2O$ and generation of stress in porous materials. Cryst. Growth Des. **8**(1), 336–343 (2008)
33. Yang, P., et al.: Hygroscopicity measurement of sodium carbonate, β-alanine and internally mixed β-alanine/Na_2CO_3 particles by ATR-FTIR. J. Environ. Sci. **87**, 250–259 (2020)

34. Sælensminde, J., Samarakoon, S.S.M.: Deterioration of concrete structures due to salt crystallization; influence of salt types and relative humidity. In: Hansen, K.K., Rode, C., Nilsson, L-O. (eds.) Proceedings on International RILEM Conference on Materials, Systems and Structures in Civil Engineering Conference segment on Moisture in Materials and Structures, pp. 120–127. RILEM Publications SARL (2016)
35. Tian, Y., et al.: Raman spectroscopy of sodium chloride under high-pressure and high-temperature. arXiv preprint arXiv:1903.11824 (2019)

Adaption of Imaging Techniques for Monitoring Cultural Heritage Objects

Amalia Siatou[1,2](✉), Athanasia Papanikolaou[3], and Evdokia Saiti[4]

[1] HES-SO, Haute Ecole Arc Conservation-Restauration, Neuchâtel, Switzerland
amalia.siatou@he-arc.ch
[2] Laboratory of Imaging and Artificial Vision, UBFC, Dijon, France
[3] Faculty of Mechatronics, WUT, Warsaw, Poland
Athanasia.Papanikolau@pw.edu.pl
[4] Department of Computer Science, NTNU, Trondheim, Norway
evdokia.saiti@ntnu.no

Abstract. The paper describes the ongoing research on an interdisciplinary approach regarding the technological developments adapted for monitoring CH objects. It covers aspects from data capturing, to data processing and cross-time registration methodologies. The work of three individual projects, that are carried out in the framework of ITN-CHANGE (Horizon 2020, GA 813789) project, are presented. These projects are based on the different backgrounds and expertise of the co authors which, when combined, can cover a wide spectrum of information indispensable for the accurate monitoring of CH objects. The potentiality of 3D Digital Image Correlation (3D DIC) for monitoring in and out of plane displacements as well as advances in Reflectance Transformation Imaging (RTI) for data processing for monitoring specular surfaces, are examined. Computational cross-time and multi-modal registration algorithms are developed for correlating 3D non-registered data over-time. Feasibility studies on mock-ups and simulated data are presented for the validation of the adapted methodologies.

Keywords: Monitoring · Cross-time registration · Reflectance transformation imaging · 3D Digital Image Correlation

1 Introduction

The interaction of Cultural Heritage (CH) objects with the environment can result in changes of their physical properties and their appearance attributes. Extensive research has been carried out on imaging methods for understanding,

This work has received funding from the European Union's Horizon 2020 research and innovation program under the Marie Skłodowska-Curie grant agreement No. 813789. A. Papanikolaou greatly appreciates the financial support granted by the Scientific Council of the Discipline Automatic, Electronics and Electrical Engineering, WUT, grant agreement No. 504/04542/1143/43.020004.

A. Osman and A. Moropoulou (Eds.): *Advanced Nondestructive and Structural Techniques for Diagnosis, Redesign and Health Monitoring for the Preservation of Cultural Heritage*, SPM 16, pp. 38–47, 2022.
https://doi.org/10.1007/978-3-031-03795-5_6

documenting and monitoring these type of changes. Imaging techniques provide powerful tools for capturing and tracking CH object alterations [1]; however, monitoring is possible only with appropriate data processing through computational methods and data interpretation by experts [2,3]. CHANGE" [4] is a research and innovation program under the auspices of the European Union's Horizon 2020 programme. The project aims at developing imaging techniques and systems for the acquisition, documentation and monitoring of CH objects. The final goal is to combine and correlate expertise of different fields under a common framework. To this sense, this paper draws upon knowledge from the fields of optical metrology, computer engineering and conservation science in an interdisciplinary approach for the development of change detection methodologies. The scientific activities are grouped into three pillars which cover different technological aspects. The first pillar refers to the strategies and systems for capturing and tracking changes on CH objects, whereas the second applies computational methods for studying cross-time changes and the last presents feasibility studies for the validation of protocols developed within the other two pillars. Currently, each pillar advances independently, with the aim to merge with the perspective to merge them as the project evolves. The remainder of the paper is organized as follows: In Sect. 2, *change capture and tracking strategies are discussed* while in Sect. 3 the problem of *data registration* is analyzed. Section 4 covers *data interpretation* and the paper concludes in Sect. 5 *with discussion and future aspects.*

2 Change Capture and Tracking Strategies

The first pillar describes technological tools and methodologies adapted for data acquisition at different time intervals using 3D DIC and RTI.

2.1 Digital Image Correlation (DIC)

DIC is a versatile, full-field, optical metrology technique with applications mainly in mechanical and civil engineering. Typically, 3D DIC is used to study heterogeneous materials under different loading conditions and to accurately calculate the maps of in and out of plane displacements and strains [5]. In plane displacements correspond to deformation in the X and Y axes, and out of plane to Z, providing thus the arbitrary 3D vector of displacements. While 3D DIC a is portable, non-invasive and low-cost technique, with an adjustable field of view, its application to CH objects can be challenging. In particular, to achieve optimum results and sub-pixel accuracy, this technique requires the surface under investigation to have a random texture that meets specific criteria (e.g. randomly positioned speckles with adequate contrast, firmly adherent pattern, etc.) [5].

A common 3D DIC set-up consists of two cameras simultaneously capturing pairs of images of the object under a certain geometrical configuration. Appropriate calibration protocols are applied to enable the triangulation and correlation of the captured data set (Fig. 1). The correlation algorithm works by detecting

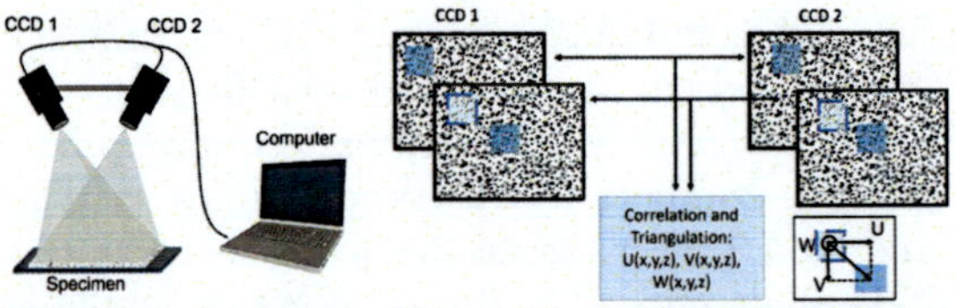

Fig. 1. Representation of (a) 3D DIC configuration and (b) working principle

intensity differences among adjacent groups of pixels (called subsets). Each subset is then localized through the subsequent images that correspond to different deformation stages, with a specific searching step. Both the subset and step values are user-defined parameters that need to be adjusted according to the object under investigation and the experimental configuration (i.e. the optical magnification, density and size of the speckle pattern). For the correlation to be feasible, each subset of the image should contain a unique pattern, thus enabling the calculation of the displacement and strains. The accuracy of the calculations and data processing time depends on the selection of subset and step size. For many CH objects, monitoring of the deformations and strains is considered necessary to understand surface alterations. Here, the 3D DIC technique can be a solution, as it can be used *in situ*, in full field-of-view (FoV) and at selected time intervals. Nevertheless, introducing an artificial speckle pattern on a CH object is not permitted. Thus, a compromise to the calculation accuracy is necessary or alternative approaches might be considered [6]. Some examples of 3D DIC applications to CH objects include displacements measurement on model canvas with random pattern [7–9], and historical parchment [10] provoked by controlled changes in the relative humidity, as well as, mechanical displacements induced on canvas paintings [6,7].

2.2 Reflectance Transformation Imaging (RTI)

RTI is a multi-light technique following a fixed configuration with a camera positioned perpendicular to an object for acquiring a set of images at different light angles (Fig. 2). This multi-angle illumination can provide photometric and geometric documentation of surfaces [11]. It has found application in CH as an easy-to-use, non-invasive, portable technique [12]. There are numerous references on the application of RTI in CH, varying in methodologies and material applications; however, most address enhancing legibility and surface details related to topography, such as examining artists' brush-strokes, or deciphering epigraphs [13,14]. In this section the feasibility of monitoring of objects is investigated. A dome with a fully calibrated light source and motorized camera functions is used to ensure the reproducibility and cross-time registration of the acquired data [15,16]. One of the system's novelties is the ability to extract raw RTI data, providing the possibility to further data processing that goes beyond the simple visualization or image enhancement [16]. In particular, geometric and statistical calculations of the stack of images produced can provide information related to the surface topography or the per-pixel reflectance response of the surface at different light angles, respectively. This results in visualization through maps depicting the surface features (features maps) that can either enhance or isolate surface information.

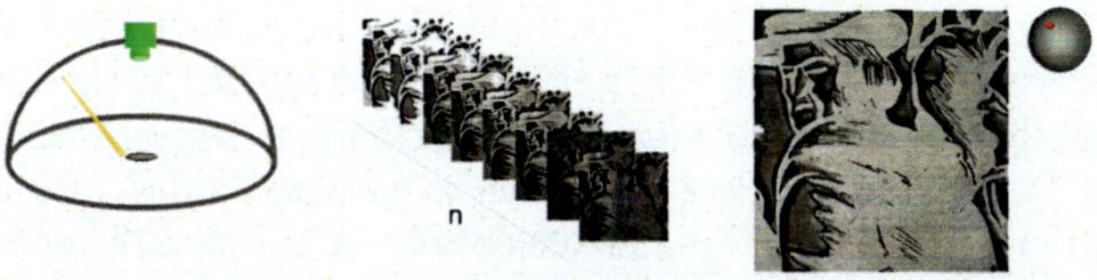

Fig. 2. Simplified representation of RTI from left to right: The dome system (left). Acquisition of a stack of images at "n" light positions (middle). Relightable images at selected light direction (right).

In this paper monitoring the condition of specular surfaces was selected. A global examination was followed, consisting of data acquisitions of artificially aged mock-ups at different time intervals. The methodology consists of data acquisitions in different RTI-modalities and correlation of the results based on visual inspection and comparison to imaging techniques routinely used in the field of CH. To ensure repeatability, the systems used and the acquisition parameters remain stable throughout the experimental process. Interpretation of results relies on CH expertise.

3 Computational Methods for Cross-Time Registration

Following the 3D model acquisition of the CH objects, the data need to be analyzed in order to accurately understand and monitor any change [17]. In general, data captured from different acquisitions can be geometrically and chronologically incoherent. In order to facilitate their study and detect changes, the data need to be registered. Registration aims to find the transformation (rotation and translation) that optimally aligns two or more instances of the same objects at different times (cross-time data), from different viewpoints (multi-view data) or by different modalities (multi-modal data) in order to bring the data in a common reference frame [18].

3.1 Cross-time Registration

Methods that monitor the geometric change of an object over time, try to compare the 3D representation of the same object captured at different time intervals. Considering that modifications may have occurred on the surface of the object (i.e. surface alteration due to weathering or conservation-restoration treatments), shape differences may have encountered between acquisitions, resulting to a non-trivial correspondence of the object's surface. Moreover, acquisition processes cannot always ensure that cross-time captures will be at the exact same position. Thus, accurate 3D spatial relations between data from different acquisitions may not be directly obtained, which makes the cross-time registration process a challenging task.

Given two 3D point clouds of the same object, but captured at different time frames, the aim of the 3D cross-time registration is to find the unknown rigid transformation so as to align the two point clouds as accurately as possible. Currently, our research is focused on the surface alterations due to weathering, where the examined object is assumed to have been uniformly exposed to weather conditions, both spatially and temporally. A framework for cross-time 3D registration is proposed in [19] that copes with big data using a down-sampling scheme that is appropriate for objects exhibiting uniform change over time. The proposed method generally outperforms the state-of-the-art in both accuracy and efficiency (Fig. 7).

4 Data Interpretation on Selected Case-Studies

The final part is the application of the above described methodologies to CH objects.

4.1 3D DIC Data Interpretation

Examples of 3D DIC application to CH objects with inherent surface texture and patterns, adequate to perform the analysis, are presented. The first case study is an oil painting on canvas that was subjected to deformation by applying mechanical pressure (loading, simply by pushing outwards) on its back surface. The pressure was applied with the intent to create a random and complex deformation distribution, in order to examine the effectiveness of the 3D DIC and its effectiveness on detecting changes using the surface features of the painting and without applying artificial texture patterns. The painting and the results of the 3D DIC study of the loading are presented in Fig. 3.

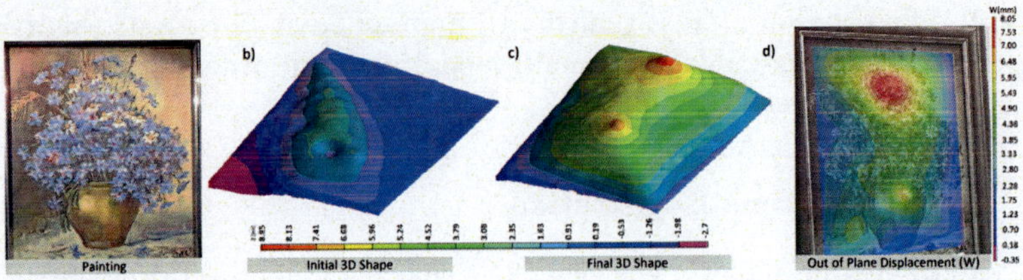

Fig. 3. 3D DIC analysis of a painting without an artificial pattern. The 3D shape maps before (b) and after loading (c), along with the 2D map out of plane displacement map (d) are presented.

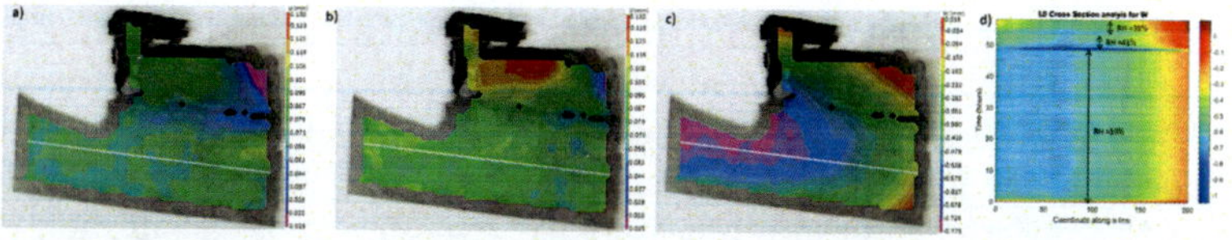

Fig. 4. 3D DIC analysis on a parchment without an applied artificial texture. The in- (U and V in a, and b respectively) and out-of-plane (W in c) displacement maps corresponding to 50% of RH are shown. In 4d the spatio-temporal analysis of W for the line corresponding to 60h duration is presented.

The second case study is the calculation of displacements in a historical parchment subjected to an environment with fluctuating Relative Humidity (RH) (Fig. 4). The in and out of plane displacements, that correspond to 50% RH, are presented along with a selected line which is analysed through time. The spatio-temporal map of the out of plane displacement (W), that corresponds to the selected line, is shown in (Fig. 4 d). During the observation time RH is gradually decreased over a duration of 60 h.

These objects represent two important groups of CH artifacts that are sensitive to environmental fluctuations and mechanical damage. For both examples, 3D DIC has provided valuable information for characterizing the surface displacements caused by different factors, detailed information can be found in [6,10]. These case studies represent short-time measurement sessions (with stable object-sensor configuration), which eliminates the need for cross-time data alignment. They, support the potential of condition monitoring of CH objects, as well as, the capability of monitoring conservation-restoration treatments with 3D DIC.

4.2 RTI Application and Interpretation

RTI was investigated for its feasibility to monitor the formation fingerprints during the tarnishing of silver objects [20]. Data were acquired for extracting information on surface appearance attributes, with the goal to isolate information related to the topography and identify characteristics related to the reflectance response. A fingerprint was placed on pure silver coupons, with an isotropic texture, and were subsequently artificially tarnished at different levels, corresponding to change over time. The surfaces were examined both with techniques routinely used for CH object documentation as well as the proposed RTI methodology [20].

The routine imaging techniques consist of calibrated photography using a light-box and correcting data through a color-checker, documentation of the surface under high magnification using optical microscopy and color measurements with a spectrophotometer.

The RTI experimental set-up involved a dome and acquisitions were preformed using a monochromatic camera; whereas, multi-light illumination was achieved using a single light source (high-power collimated white LED light).

For each acquisition set, 150 light positions were acquired covering an azimuth angle from 0° to 360° and an elevation angle of 5°–60°. All acquisitions were homogeneous, i.e., the lighting positions were spread uniformly around the dome covering an entire hemisphere and thus providing overall angular illumination of the surface. To ensure repeatability, exposure time, acquisition parameters and selected ROI (region of interest) were kept constant for the different tarnish levels. Data processing consisted of calculating the per-pixel mean reflectance response of the stack of images and visualising the results through gray-scale colormaps.

Figure 5 presents an example of the comparison between different imaging techniques for registering information related to monitoring the cross-time surface change of fingerprints on silver. Despite the different scales presented for each technique, the possibility of enhancing or isolating specific information related to the change of the reflectance response of the surface, in the area of the fingerprint, is evident in the mean reflectance response of the RTI data, in a form of gray-scale colormap, even at light levels of tarnish.

Global examination of feature maps, at different levels of silver tarnishing in the presence of fingerprints, has shown promising results in the ability to detect and enhance visualization in comparison to routinely used imaging techniques or usual RTI visualization and surface enhancement. Furthermore, from the CH perspective, the detection of fingerprints at low levels of tarnish, which is difficult to document through regular inspection, provides a tool for non-invasive examination of CH surfaces. However, for quantification of results, further data processing is necessary to better evaluate surface change over time and to apply actual cross-time registration on the examined surfaces.

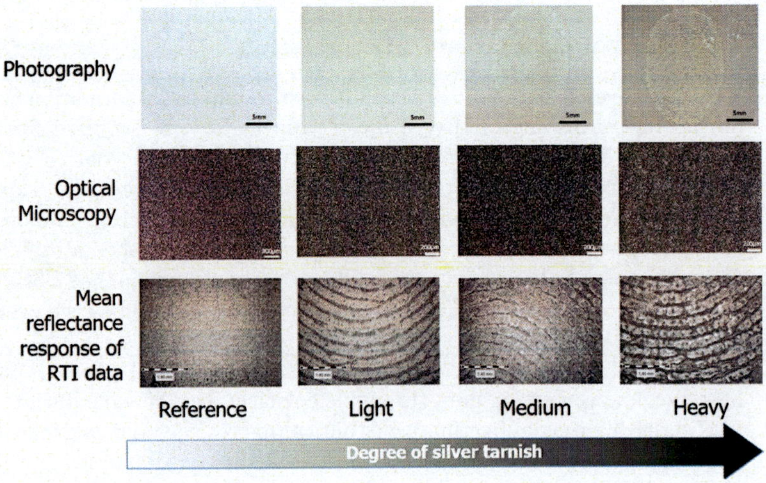

Fig. 5. Monitoring silver tarnishing by RTI features maps and comparison to the routinely used imaging techniques of photography and optical microscopy.

4.3 Cross-time Registration Data Interpretation

The main challenge in cross-time registration is the lack of a publicly available dataset with altered objects with the respective ground truth. In order to overcome this, a dataset of weathered CH objects was synthetically created. Starting from the publicly available dataset [21], first a random rotation and translation were applied to the objects; then two weathering effects were simulated and applied on the transformed objects to create the relative surface alteration. The effects simulated are the dry deposition of crust due to pollution and the recession by acid rain. These effects can result in gain or loss of material on the surface of an object.

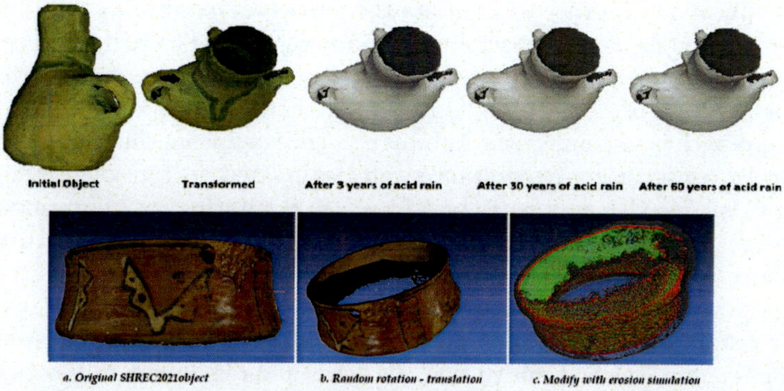

Fig. 6. Up: The steps of the dataset creation for one object. The object is initially transformed and then the erosion simulator runs for 20 epochs of 3 years each. In this example, the initial model is shown degraded due to the effect of acid rain after 3, 30 and 60 years. Down:Simulated dataset for weathering. Original CH object from SHREC2021 dataset (a), along with the transformed instance (b). On (c), different weathered data are depicted. The reference object is depicted in gray color, the object after 30 years of ageing in red and the after 60 years in green.

Since weathering is performed *in situ* and the transformation parameters are known, the ground truth for benchmarking cross-time registration algorithms can be acquired. The process is outlined in Fig. 6. The training part of the dataset is then used to train our deep network to register weathered objects. The proposed method (Fig. 7) first down-samples the reference and weathered point clouds using their principal curvatures. Then, the down-sampled point clouds are segmented and finally, the registration is performed by aligning the component centroids of each segment.

5 Discussion and Future Aspects

In summary, different imaging techniques and methodologies are being examined and adjusted by the authors to meet the requirements for monitoring CH objects

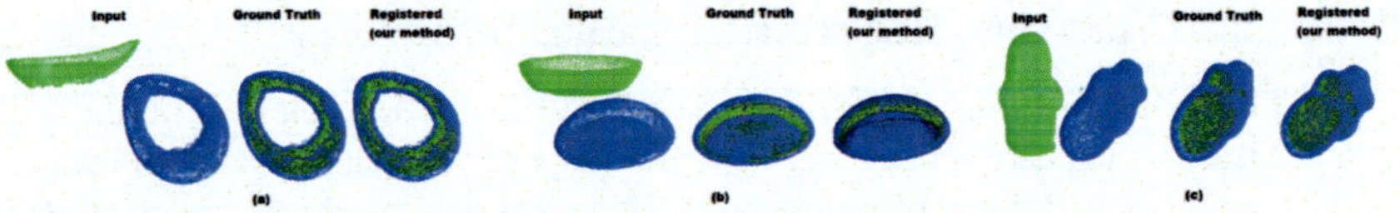

Fig. 7. Examples of the simulated weathered dataset along with the results of the proposed methodology for cross-time registration.

in terms of set-ups, proper registration and data processing. For the accurate interpretation of the results with the developed methodologies, a key factor is proper data assessment in collaboration with CH specialists. As a future step, we consider the combination of the presented methodologies in an end-to-end application, in order to facilitate the process of monitoring CH objects.

Through the interdisciplinary approach of the ITN-CHANGE project different perspectives in terms of technologies and expertise are contributing in the adaptation of new methodologies for monitoring CH objects. The aim of this approach is to connect the information collected from the aforementioned pillars with conservation-restoration strategies related to the long-term preservation of tangible CH with particular focus on monitoring, by capturing and tracking changes. Currently, each pillar is developed independently with the intent to later combine the processes and information under a common framework. To achieve the interdisciplinary goals of the CHANGE project, researchers are in constant dialogue, collaboration and training to familiarize themselves with the different scientific fields so that the needs and requirements of each specialty can be met and understood. The cooperation and teamwork of scientists with different backgrounds is considered a necessary link for the adaptation of technologies from different fields to CH.

References

1. MacDonald, L.: Digital Heritage: Applying Digital Imaging to Cultural Heritage. Routledge, Milton Park (2006)
2. Pintus, R., et al.: Geometric analysis in cultural heritage. In: GCH, pp. 117–133 (2014)
3. Saha, S., Siatou, A., Sitnik, R.: Classification of surface geometry behavior of cultural heritage surfaces based on monitoring change. In: Optics for Arts, Architecture, and Archaeology VIII. vol. 11784. International Society for Optics and Photonics (2021)
4. CHANGE cultural heritage analysis for new generations, European union's horizon 2020 research and innovation programme. https://change-itn.eu/. Accessed Apr 2021
5. Sutton, M.A., Orteu, J.J., Schreier, H.W.: Image correlation for shape, motion and deformation measurements : basic concepts, theory and applications. Springer, Boston (2009). https://hal-mines-albi.archives-ouvertes.fr/hal-01729219, https://doi.org/10.1007/978-0-387-78747-3
6. Papanikolaou, A., Garbat, P., Kujawinska, M.: Colour digital image correlation method for monitoring of cultural heritage objects with natural texture. In: Liang, H., Groves, R. (eds.) Optics for Arts, Architecture, and Archaeology VIII, vol.

11784, pp. 166–177. International Society for Optics and Photonics, SPIE (2021). https://doi.org/10.1117/12.2592549

7. Kujawinska, M., et al.: Digital image correlation method: a versatile tool for engineering and art structures investigations. In: Rodríguez-Vera, R., et al. (eds.) 22nd Congress of the International Commission for Optics: Light for the Development of the World, vol. 8011, pp. 2599–2606. SPIE (2011). https://doi.org/10.1117/12.915566
8. Malesa, M., et al.: Application of digital image correlation for tracking deformations of paintings on canvas. In: Pezzati, L., Salimbeni, R. (eds.) O3A: Optics for Arts, Architecture, and Archaeology III, vol. 8084, pp. 157–164. International Society for Optics and Photonics, SPIE (2011). https://doi.org/10.1117/12.889452
9. Malowany, K., et al.: Application of 3d digital image correlation to track displacements and strains of canvas paintings exposed to relative humidity changes. Appl. Opt. **53**(9), 1739–1749 (2014). https://doi.org/10.1364/AO.53.001739
10. Papanikolaou, A., Dzik-kruszelnicka, D., Saha, S., Kujawinska, M.: 3D digital image correlation system for monitoring of changes induced by RH fluctuations on parchment. In: Proceedings of the IS&T International Symposium on Electronic Imaging: 3D Imaging and Applications, pp 65-1–65-7 (2021). https://doi.org/10.2352/ISSN.2470-1173.2021.18.3DIA-065
11. Castro, Y., et al.: A new method for calibration of the spatial distribution of light positions in free-form RTI acquisitions. In: SPIE Optical Metrology, 2019, Munich, Germany, vol. 11058, p. 38. SPIE, Munich, Germany, June 2019. https://doi.org/10.1117/12.2527504, https://hal-univ-bourgogne.archives-ouvertes.fr/hal-02353517
12. CHI 2021: Cultural heritage imaging. http://culturalheritageimaging.org/Technologies/RTI/ Accessed Jun 2021
13. Mudge, M., et al.: Image-based empirical information acquisition, scientific reliability, and long-term digital preservation for the natural sciences and cultural heritage. Eurographics (Tutorials) **2**(4) (2008)
14. Earl, G., et al.: Reflectance transformation imaging systems for ancient documentary artefacts. Electron. Vis. Arts (EVA 2011) 147–154 (2011)
15. Pitard, G., et al.: Discrete modal decomposition: a new approach for the reflectance modeling and rendering of real surfaces. Mach. Vis. Appl. **28** (2017). https://doi.org/10.1007/s00138-017-0856-0
16. Nurit, M., et al.: HD-RTI: An adaptive multi-light imaging approach for the quality assessment of manufactured surfaces. Comput. Ind. **132** (2021). https://doi.org/10.1016/j.compind.2021.103500
17. Saha, S., Foryś, P., Martusewicz, J., Sitnik, R.: Approach to analysis the surface geometry change in cultural heritage objects. In: El Moataz, A., Mammass, D., Mansouri, A., Nouboud, F. (eds.) ICISP 2020. LNCS, vol. 12119, pp. 3–13. Springer, Cham (2020). https://doi.org/10.1007/978-3-030-51935-3_1
18. Saiti, E., Theoharis, T.: An application independent review of multimodal 3D registration methods. Comput. Graph. **91**, 153–178 (2020)
19. Saiti, E., Danelakis, A., Theoharis, T.: Cross-time registration of 3d point clouds. Comput. Graph. **99**, 139–152 (2021)
20. Siatou, A., et al.: Surface appearance assessment as a tool for characterizing silver tarnishing, November 2020. https://doi.org/10.5281/zenodo.4299912
21. Sipiran, I., et al.: Shrec 2021: retrieval of cultural heritage objects. Comput. Graph. **100**, 1–20 (2021)

Documenting Artifacts Using 3D Representation and Nondestructive Evaluation Tools

D. A. Exarchos[1(✉)], S. G. Farmaki[1], I. K. Tragazikis[1], A. C. Mpalaskas[1], A. Vasios[2], V. Papadopoulou[2], and T. E. Matikas[1]

[1] Department of Materials Science and Engineering, University of Ioannina, 45110 Ioannina, Greece
d.exarchos@uoi.gr

[2] Ephorate of Antiquities of Arta, Greek Ministry of Culture and Sports, 47100 Arta, Greece

Abstract. The sustainability of antiquities is a very difficult and complicated task. It would be of great importance the creation of a database containing fusion data not only on the condition and detailed and accurate representation of the artifacts but also about the restoration and preservation provided by quantitative nondestructive techniques such as Infrared Thermography. The 3D model with all the related metadata would be representing a digital identity of the artifact, therefore guaranteeing the sustainability of the item. Additionally, it will be a useful tool towards the validation/authentication of the artifact in case it is damaged or missing.

Keywords: 3D scanning · Nondestructive testing · Fusion data · Infrared Thermography

1 Introduction

The rapid progress of different 3D scanning methods has been aided by technical advancements in recent years. Many economic and industrial activities, such as entertainment, industrial design and medicine, benefit from the accuracy and speed. Using this property large and small items can be scanned. However, the cultural heritage sector has arguably reaped the largest benefits [1–7]. In 1980s, archaeologists were able to document and map archaeological sites and items at a scale and accuracy never before imaginable because of the introduction of three-dimensional (3D) surface modeling. This technology is widely used in the domains of archaeology, paleoanthropology and heritage preservation [8–16]. Cultural heritage is testimonial to people's communities and cultural groups' existence, identity, and viability and it represents the fundamental base of societies all over the world. Cultural heritage protection, preservation and promotion are a global obligation for future generations to pass down historical memory.

The main element for the preservation of architectural heritage is the thorough documentation and recording of traditional and cultural properties using various techniques such as traditional geodetic measurements, photogrammetry, and terrestrial scanning

A. Osman and A. Moropoulou (Eds.): *Advanced Nondestructive and Structural Techniques for Diagnosis, Redesign and Health Monitoring for the Preservation of Cultural Heritage*, SPM 16, pp. 48–54, 2022.
https://doi.org/10.1007/978-3-031-03795-5_10

with lasers, which are converted into two-dimensional and three-dimensional models with photorealistic capabilities. As cultural heritage monuments are damaged and finally destroyed by environmental causes such as natural catastrophes, vandalism, accidents or the natural aging of the building blocks, up-to-date and accurate digitization becomes increasingly necessary [17–20]. In order the cultural heritage to be sustainable, each monument must be accurately evaluated for its uniqueness and significance. A sustainable method for the protection of cultural heritage can be realized using virtual reality and the production of a realistic 3D model provides information not only about the materials but also about their condition. As a result, could be complex deciding which system or hardware to be utilized, especially since many of these systems are created and tested in (and for) laboratory environments.

Today, there are a variety of 3D laser scanners which provide advanced solutions on 3D digitations [5]. 3D scanners can be classified into contact and non-contact 3D scanners. Due to the delicate nature of the artifacts, contact scanners are not often used in the field of cultural heritage. Non-contact scanners, on the other hand, have been employed successfully in cultural digitization initiatives for the previous decade. Structured light, structure-from-motion, and computer tomographic (CT) rendering are three 3D surface techniques that are well-represented in academia. The most common technique is based on the principle of projection some form of structured or patterned light onto the object being scanned [21, 22]. After that, the scanner uses this known structure or pattern to figure out the object's 3D shape [23].

Structure from Motion (SfM) is a technique that based on the theory of stereoscopic photogrammetry, which claims that overlapping offset photos can be used to derive a 3D geometry. The SfM method varies from prior systems, in that the software automatically determines the object's coordinates, as well as the location and angle of the cameras. As the camera position and angle vary, matching features in each set of images are tracked. The software uses a point-based method to determine the nature of the features and how these points change in successive images. As a result, a complete 3D model can be created without the necessity of pre-determined coordinates or camera location. Furthermore, the larger the degree of overlap between each image collection, the higher the produced 3D model's quality and accuracy. SfM technique, like structured light technique, uses a variety of approaches and procedures, as well as ways that combine the two systems, resulting in a large number of alternative systems and hardware to pick from [24–26].

In addition, monitoring the structural health of cultural heritage monuments via quick and reliable nondestructive assessment techniques is of great importance [27, 28]. One of the methods which is Infrared Thermography (IRT). IRT is a powerful nondestructive evaluation tool which can be effectively used for defect detection in materials such as ceramic composites. It is a full field, non-contact, and real time method. One major advantage of thermography applications is the detection of moisture and monitoring of sub-surface cracking which provide important information for the preservation and conservation treatments [29]. IRT approaches are divided into passive and active ones, depending on thermal stimulation requirements. In passive IRT no external thermal excitation is needed for defect inspection while in active IRT external thermal stimulation is applied in a uniform way to induce thermal contrasts [30]. Examples of active thermography heat sources are flash lamp, quartz lamp, eddy current, microwave and ultrasound

[31]. Depending on the external source of stimulation the active thermography is categorized to: Step heating (SH), vibrothermography (VT), Pulse Thermography (PT), Pulse Phase Thermography (PPT) and Lock-in Thermography (LT) [32].

In this study Optical Lock-in Thermography (LT) was used. LT is based on thermal waves generated inside the specimen by a periodic sinusoidal thermal stimulation. The thermal radiation emission captured by a thermographic camera, converted by the Fourier transform (FFT) and provide thermal images (thermograms).

2 Experimental Study

2.1 Experimental Set-Up

In the current study two different type of 3D scanners have been used. For larger artifacts (200 to 400 mm height) white light (LEO) scanner has been used while smaller artifacts (50 to 200 mm height) have been scanned by blue light scanner. The scanner measures the variations from different perspectives and then creates the 3D geometry with high precision due to light projection at predetermined intervals and frequencies. The sensitivity of this system to sunlight has a significant impact on the quality and accuracy of the 3D produced object. Similarly, these scanners require an external power source to operate, restricting even further the spectrum of possible field applications in archaeology. However, the new generation of scanners, such as LEO, has eliminated need of external power supply (Fig. 1).

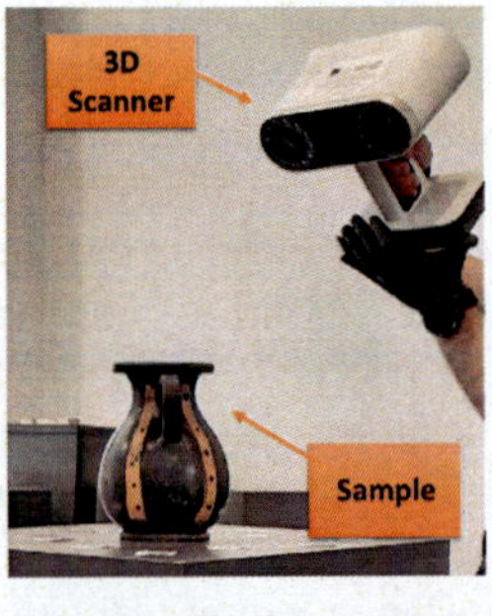

Fig. 1. Experimental set-up of 3D scanner.

For the realization of experiments were used: an IR sensor (operating in the longwave infrared (LWIR) band and resolution 640 × 480 pixels). The thermal excitation source which selected for continuous and uniform heat flux, was a halogen lamp power of 300 W. The IR sensor was placed at the distance of 40 cm from the sample. The IR frequency was 0.01 Hz, and the halogen lamp amplitude was 90% (Fig. 2).

2.2 Results and Discussion

Several artifacts for which the Ephorate of Antiquities of Arta, Greece has given access, have been scanned with the use of the aforementioned, state of the art, 3D scanners and

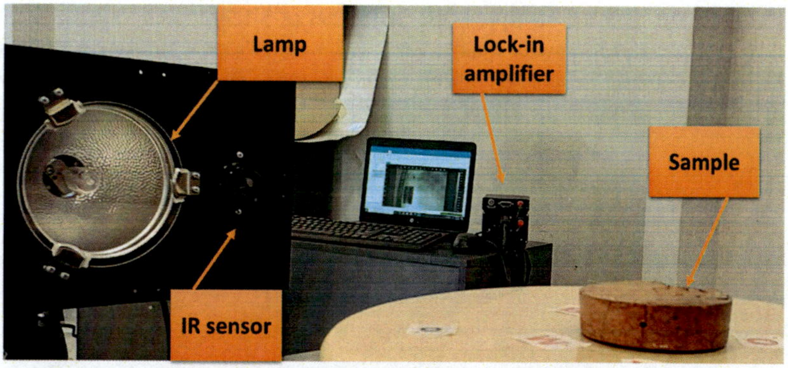

Fig. 2. Experimental set-up of Optical Lock-In IRT.

also the items have been assessed with the use of Optical Lock-in Infrared thermography. All the data as well as data about the documentation that were provided by the Ephorate of Antiquities have been fused in order to create a unique 3D fusion data base which will be the "Identity" of the artifact and in this way, will provide a quick, accurate and complete overview of the item itself and also the condition regarding the conservation and their structural health.

In Fig. 3 it can be observed the artifact 3D model which is very detailed and has a below millimeter accuracy. Also, for each side of the 3D model several high quality and resolution images are taken and also the data of the IRT are also oriented and scaled in order to fit the model and the photographs. The IR thermographs reveal that at the subsurface of the artifact exist areas that need to be investigated in order to be clear if these areas have delamination and if further restoration is need.

Fig. 3. 3D model of an artifact and respectively the photograph and the fused photograph with part of the thermograph

In Fig. 4 different aspects the 3D model of a red-figured "hydria" is shown and also in the second row of images the fused model with the IRT data is shown. Infrared thermography reveals useful information of the structural integrity of the "hydria" and in addition it reveals details of the gestures that are not visible with naked eye.

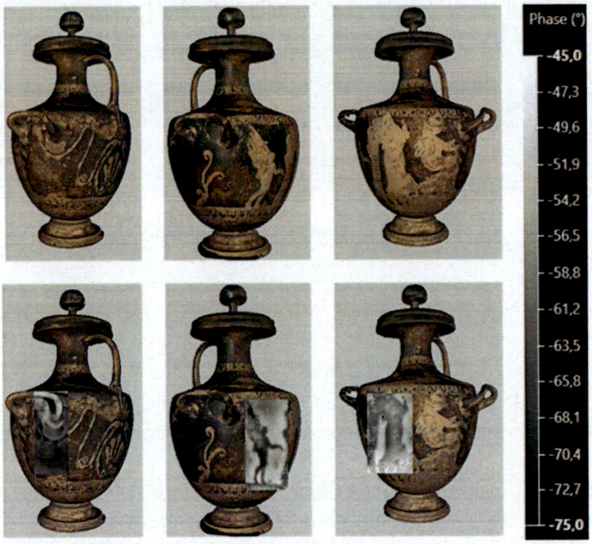

Fig. 4. 3D model of a burial "Hydria" rotated in different angles and also the fused model with data from the IRT

This database with many artifacts and different multidisciplinary data will be updated each time something is changed to the artifacts, either due to conservation work or due to an unexpected event. In this way, the database will also contain the entire history of the items from the moment they were discovered until the day a change occurs. Such a tool will be available in a very friendly environment of a digital visualized database which will be useful not only for the conservators but to the archeologists and the community in general.

Acknowledgements. This work was co-financed by Greece and the European Commission under the "National Strategic Reference Framework" (2014–2020), Special Action "Open Innovation in Culture", project title: "Development of innovative methodology for nondestructive evaluation and the promotion of cultural heritage monuments (INSPIRE)" (Project reference number: 5048509).

References

1. Soile, S., et al.: Multidisciplinary documentation: combination of NDTs with geometric documentation for the diagnostic study of Ancient Temples-The case of the Pythian Apollo Temple in Rhodes, p. 18 (2019)

2. Delegou, E.T., et al.: The effect of fire on building materials: the case-study of the Varnakova Monastery Cells in central Greece **2**(2), 1233–1259 (2019)
3. Ioannidis, C., Tsakiri, M.: Laser scanning and photogrammetry for the documentation of a large statue-experiences in the combined use. In: Proceedings of CIPA XIX International Symposium (2003)
4. Kan, T., et al.: Rapid 3D digitalization of the cultural heritage: a case study on Istanbul suleymaniye social complex (2019)
5. Patias, P., et al.: 3D mapping of cultural heritage: special problems and best practices in extreme case-studies **5**, W1 (2013)
6. Mwangangi, K.K.: 3D building modelling using dense point clouds from UAV. University of Twente (2019)
7. Hatzopoulos, J.N., et al.: Use of various surveying technologies to 3D digital mapping and modelling of cultural heritage structures for maintenance and restoration purposes the tholos in Delphi, Greece **17**(3) (2017)
8. Rüther, H., et al.: Laser scanning for conservation and research of African cultural heritage sites: the case study of Wonderwerk Cave, South Africa. J. Archaeol. Sci. **36**(9), 1847–1856 (2009)
9. Armstrong, B.J., et al.: Terrestrial laser scanning and photogrammetry techniques for documenting fossil-bearing palaeokarst with an example from the Drimolen Palaeocave System, South Africa **25**(1), 45–58 (2018)
10. Ashley, K., Patrick, R.-Q., Marina, E.: Multimodal spatial mapping and visualisation of Dinaledi Chamber and Rising Star Cave. South African J. Sci. **112**(5/6) (2016)
11. Subsol, G., et al.: In situ 3D digitization of the "Little Foot" australopithecus skeleton from sterkfontein. PaleoAnthropology, 44–53 (2015)
12. Siebke, I., et al.: The application of different 3D-scan-systems and photogrammetry at an excavation—A Neolithic dolmen from Switzerland. Digit. Appl. Archaeol. Cult. Heritage **10**, e00078 (2018)
13. Clarkson, C.: Measuring core reduction using 3D flake scar density: a test case of changing core reduction at Klasies River Mouth, South Africa. J. Archaeol. Sci. **40**(12), 4348–4357 (2013)
14. Roman, C., Inglis, G., Rutter, J.: Application of structured light imaging for high resolution mapping of underwater archaeological sites. In: OCEANS 2010 IEEE Sydney (2010)
15. McPherron, S.P., Gernat, T., Hublin, J.-J.: Structured light scanning for high-resolution documentation of in situ archaeological finds. J. Archaeol. Sci. **36**(1), 19–24 (2009)
16. Baier, W., Rando, C.: Developing the use of Structure-from-Motion in mass grave documentation. Forensic Sci. Int. **261**, 19–25 (2016)
17. Verhagen, P., Nuninger, L., Groenhuijzen, M.R.: Modelling of pathways and movement networks in archaeology: an overview of current approaches. In: Verhagen, P., Joyce, J., Groenhuijzen, M. (eds.) Finding the Limits of the Limes. Computational Social Sciences(), pp. 217–249. Springer, Cham (2019). https://doi.org/10.1007/978-3-030-04576-0_11
18. Sofocleous, E., et al.: The geometric documentation of the asinou church in Cyprus (2006)
19. Fassi, F., et al.: Comparison between laser scanning and automated 3D modelling techniques to reconstruct complex and extensive cultural heritage areas **5**, W1 (2013)
20. Pavlidis, G., et al.: Methods for 3D digitization of cultural heritage **8**(1), 93–98 (2007)
21. Bell, T., Li, B., Zhang, S.: Structured light techniques and applications. In: Wiley Encyclopedia of Electrical and Electronics Engineering, pp. 1–24 (1999)
22. Tucci, G., et al.: High-quality 3D models and their USE. Int. Arch. Photogramm. Remote Sens. Spatial Inf. Sci. **XLII-2/W5**, 687–693 (2017)
23. Cooper, S.D., et al.: Examination of the potential of terrestrial laser scanning and structure-from-motion photogrammetry for rapid nondestructive field measurement of grass biomass **9**(6), 531 (2017)

24. Alsadik, B.S.: Guided close range photogrammetry for 3D modelling of cultural heritage sites (2014)
25. Rönnholm, P., et al.: Integration of laser scanning and photogrammetry **36**(3/W52), 355–362 (2007)
26. Bolognesi, M., et al.: Testing the low-cost rpas potential in 3D cultural heritage reconstruction (2015)
27. Kordatos, E.Z., et al.: Infrared thermographic inspection of murals and characterization of degradation in historic monuments. Constr. Build. Mater. **48**, 1261–1265 (2013)
28. Kordatos, E.Z., et al.: Rapid evaluation of the fatigue limit in composites using infrared lock-in thermography and acoustic emission. Mech. Res. Commun. **54**, 14–20 (2013)
29. Kordatos, E.Z., Aggelis, D.G., Matikas, T.E.: Monitoring mechanical damage in structural materials using complimentary NDE techniques based on thermography and acoustic emission. Compos. B Eng. **43**(6), 2676–2686 (2012)
30. Farmaki, S., et al.: A novel infrared thermography sensing approach for rapid, quantitative assessment of damage in aircraft composites. Sensors (Switzerland) **20**(15), 1–16 (2020)
31. Zalameda, J., Winfree, W.: Detection and characterization of damage in quasi-static loaded composite structures using passive thermography **18**(10), 3562 (2018)
32. Maldague, X.: Introduction to NDT by Active Infrared Thermography1 (1995)

The Usefulness of X-ray Fluorescence Analysis on Photocopies Identification – A Case Study

Lydia Tsoka[1](✉), Georgios P. Mastrotheodoros[1,2], and Konstantinos Choulis[1]

[1] Department of Conservation of Antiquities and Works of Art, University of West Attica, Agiou Spyridonos 28, Egaleo, Athens, Greece
lydiatsk@gmail.com

[2] Institute of Nanoscience and Nanotechnology, NCSR "Demokritos", Patr. Gregoriou E & 27 Neapoleos Street, 15341 Agia Paraskevi, Greece

Abstract. The research was conducted in order to interpret the different visual characteristics of the photocopies and to identify materials and techniques used to properly evaluate the diagnostic stage prior to conservation. This study is an integral part of L. Tsoka's dissertation on the Postgraduate Program "Conservation of Cultural Heritage" of the Department of Conservation of Antiquities and Works of Art of the University of West Attica. Specifically, using a portable XRF device, areas of different tones (white, gray, black) were examined in six of the 316 paper sheets of which the Rena Papaspyrou's artwork "Photocopies Directly from Matter 1980–81" located at National Museum of Modern Art in Greece consists. The sheets cover the whole range of paper qualities distinguished in the artwork. Thus in the entirety of the artwork 11 sheets are uncoated high in **calcium** content typewriter products and 305 are photocopies. The 301 of the photocopies have produced with the direct electrostatic process, while the rest produced with the indirect one. This is evidenced by the **zinc** coating they carry and the high concentration of iron in the toner. From the rest, one sheet present **titanium** coating while silver is detected in the toner. The two remaining sheets are uncoated and present high concentration in calcium (but lower compared to the typewriter product), as well as **gypsum** and **aluminosilicate** content is detected. These are the only sheets with **potassium** presence. The use of photocopying techniques from contemporary artists is highlighted, as well as the advantages of the XRF technique as a method of exploring contemporary copies on paper. In fact, based on the bibliography, it is established that this is the first time that a detailed investigation of the specific type under consideration is attempted not only in Greece, while abroad research is extremely limited.

Keywords: Photocopies · X-ray fluorescence · Direct electrostatic process · Indirect electrostatic process · Filler · Coating · Toner

1 Introduction

Photocopiers have flooded the global industry since the first automatic machine - the Xerox 914 - launched in 1959 by Haloid Company (later renamed Xerox Corporation)

A. Osman and A. Moropoulou (Eds.): *Advanced Nondestructive and Structural Techniques for Diagnosis, Redesign and Health Monitoring for the Preservation of Cultural Heritage*, SPM 16, pp. 55–70, 2022.
https://doi.org/10.1007/978-3-031-03795-5_11

[1]. Offices and printing centers immediately exploit the more accessible, cheaper and quicker way of making copies supplanting the other technologies existed until then. As it was expected this innovation immediately attracted the attention of artists worldwide, making photocopying one of the first methods bridging the gap between technology and art. Just since 1960 they had begun to experiment with the creative potential of that tool developing a new art form. The artists who participated in the International Mail Art Movement[1] were the first to use xerography [2]. In 1982 Louise Neaderland, an internationally acclaimed xerox artist founded the International Society of Copier Artists in New York. She also published I.S.C.A. Quarterly, a magazine with Copy Art anthologies, in order contact within the members be maintained [3].

I.S.C.A. quarterly, whose members were mostly women, tried, in addition to promoting the medium in general, to bypass the male-dominated art world, which existed until then [4]. However, the daily accessibility of women in the use of photocopiers at the offices had the natural consequence of their utilization as a tool of art production [5]. Thus women are at the forefront of the xerography movement, as it is a means of subverting the binary creator-copyist, in whose name the work of women was underestimated, both as artists and as professionals [4]. Linda Nochlin's 1971 article entitled "Why Have There Been No Great Women Artists?" [6] placed the limits of women on artistic success and was the impetus for the beginning of feminist art. In the USA, most of the artists who use xerography in their work are women, while in other countries mainly men are active [5].

The use of the same technology by a multitude of artists does not mean that they share a common style or aesthetic. For example American Edward Meneeley (1927–2012) at the early 60 s investigated the different perspective that this new technique gave to objects by using waste (such as duct tape, typewriter ribbons, etc.) in an abstract layout and photocopied them on colorful papers [7].

In the most recent years, Swiss Dominique Teufen (1975-) uses random materials, such as plastic and paper wrappers, food packaging, flour and fabric, to represent the earth, water and sky, creating enchanting landscapes [8].

Rena's Papaspyrou (1938-) work "Photocopies Directly from Matter 1980–1981" (Fig. 1), compiled by hers wander around the city, collecting pieces/objects/materials which capture the changes of the urban landscape, the passage of time and the marks left by human intervention, <<episodes>> as she calls them. Besides, the urban areas always haunted her and inspired her. She takes advantage of the neglected elements of the cityscape and she produces art. In this project she has used, amongst others, papers with colour residues, de-walled walls, pieces of wood, cans, cloths, leafs, crumpled papers as well as two body parts, the head and the hand. She used the photocopier existed in the bailiff's office of the Athens School of Fine Arts placing the materials straight onto the photocopier screen and then photocopied them [9, 10].

[1] Mail Art is a global cultural movement that began in the early 1960s. It includes the sending of visual arts (but also music, poetry, etc.) through the international postal system. The basic principle is exchange and equal cooperation (one to one).

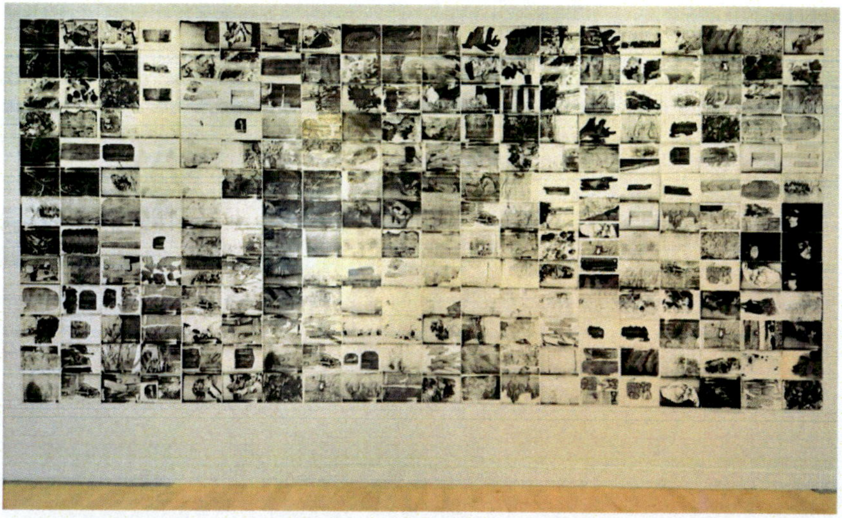

Fig. 1. The installation "Photocopies Directly from Matter 1980–81" at its display at Benaki Museum during the exhibition 'The Same River Twice' (21/06/2019–22/09/2019) (http://lizathenes.canalblog.com/archives/2019/07/02/37458058.html)

In 2014, Papaspyrou donated her work to the National Museum of Modern Art, Athens Greece. At its final form the artwork consists of 316 pieces, 305 of which are photocopies and 11 are contents pages produced with typewriter, which in the archived form of the installation, are inserted and separate the illustrated themes of the photocopied sheets. At every new exhibition the installation is displayed, Papaspyrou chooses her preferable sheets and creates a unique composition/image each time.

The 301 of the photocopies were created in the same machine which then replaced due to a technical damage. Papaspyrou examined the perspective of the new machine and the new available paper quality that came with it but the outcome wasn't satisfying her. That prevented her from continuing photocopying, however she included at her artwork those three last attempts. They are easily recognizable due to their matt texture in contrast with the light gloss the other sheets have. That glossiness raised the suspicion, during visual observation, that the copies might bear a coating.

X-Ray fluorescence has already widely used for paper characterization and identification of coating presence. However according to the bibliography for the investigation of photocopies mostly FTIR has been used, mainly at researches on the field of forensic sciences [11], making the present study quite innovative. Especially in Greece, this kind of technique has never been used before for examining contemporary artworks on paper.

2 Photocopy Processes

Electrophotography[2] [12], the combination of photoconductivity (the ability of a material to transmit electricity better after being exposed to illumination) with electrostatics (two opposite electrically charged materials attract each other) is the method Carlson Chester discovered in 1938 in order to produce the first xerographic copy (Fig. 2).

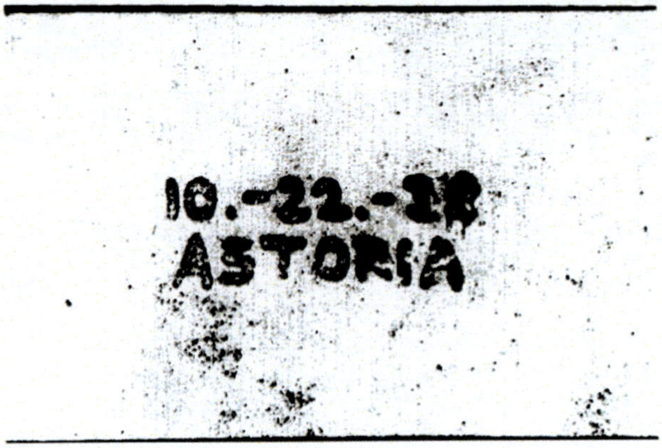

Fig. 2. The first historically xerography that states the date and place of its construction "10-22-38 ASTORIA". (Gundlach, 1989:5 [13]).

The fundamental principles that Chester set have been used since then in all the photocopiers. The revolution that his method (indirect/transfer electrostatic process as it is called) introduced is the use of plain (uncoated) paper and it's the most common photocopying process. Nevertheless there are copiers that are based on the direct electrostatic process in which coated paper is used. These types of copiers were first used in Greece, until company P. Solomos & Co introduces the first indirect electrostatic machines at 1965 [14]. Further distinction of these two methods relies in the physical state of the toner, wet or dry.

2.1 Indirect Electrostatic Process

An electrostatic charge is evenly distributed on the surface of a cylindrical metal drum or belt [15] (photoconductor) coated with a photoconductive material, most commonly amorphous **selenium** (a-Se) [16] or hydrogenated amorphous **silicon** (a-Si: H) [1] via a corona unit (corotron), ionizing the surrounding air and consequently charging the photoconductor [17]. By reflecting light from the document, through systems of lenses and mirrors, selected areas of the photoconductor are discharged. The pattern created by the charged and uncharged areas corresponds to the latent image of the original [18]. For

[2] The word "electrophotography" was replaced by the word "xerography" for commercial purposes, a term coined by an Ohio classical language teacher. It consists of the Greek words for "dry" – "ξηρός" and "writing" – "γραφή".

the development [19] of the latent to a visible image opposite polarity toner particles are deposited to the photoconductor who then comes in direct contact with the paper sheet. A corotron system charges the back of the paper and the strong electrostatic field that being developed repels the toner from the photoreceptor and forces it to be transferred to the surface of the paper. Finally, the toner image that is held by loose electrostatic forces is fixed to the sheet using heat, pressure or a combination thereof, in order to a permanent copy be created [20].

The toner particles are usually copolymers of **styrene** with **methacrylate esters** containing at a rate of 5–10% the desired pigment, in the size of a micrometer · carbon black for black toners and mixtures of dyes and pigments for colored toners [16]. At the wet indirect process the toner has the form of suspension while in the direct process it is a powder.

2.2 Direct Electrostatic Process

This type of photocopying is rare because it requires a specially coated zinc oxide (ZnO) paper. Starch or latex is used as a binder, while other organic compounds are included as additives. The methodology used to develop and fuse the direct electrostatic image is the same as that of the indirect process, except that the coated paper acts as the photoreceptor. The paper surface is uniformly charged via a corona device, either positively or negatively. The light reflected (with a system of lenses and mirrors) from the original document, breaks the charge in the areas of the background, leaving an electrostatic pattern (latent image) on the paper overlay. The development of the image is carried out by dry or wet method [21].

In this case the dry toner consists of a mixture of resin and iron particles, while the wet toner is a suspension mixture of pigment particles and solvent. The high concentration of iron as well as of zinc (coating) can easily detected by X-ray microprobe analysis differentiating this process by the indirect one [22].

3 Fillers and Coatings

At the twentieth century, wood pulp has already substituted other lignocellulosic materials (e.g. rags) for paper making. Following a mechanical, chemical, semi-chemical or chemi-mechanical process the lignin and hemicelluloses are precipitated in order to prevent deterioration over time, while cellulose is reformed into a web producing the final product [23]. The pulp may be then bleached [24] (with chlorine and its derivatives) producing white paper while inorganic particulate minerals [25] are added for the improvement of optical and qualitative properties. The particle size of the pigment powders fluctuates between 2–10 μm. Smoother and finer ones are used in paper coatings[3] whereas fillers are constituted of more rough and large particles. The most common minerals are calcium carbonate (ground-GCC or precipitated-PCC), clay, titanium dioxide and talc (Fig. 3). Among others zinc oxide, calcium sulfate (gypsum) and silica products have also being used [27, 28].

[3] The coating fills in the empty pores and spaces between fibers creating a smooth surface. This normalization improves printability as it contributes to the uniform distribution of the toner. With the contained pigments the desired gloss, brightness and opacity is achieved [26].

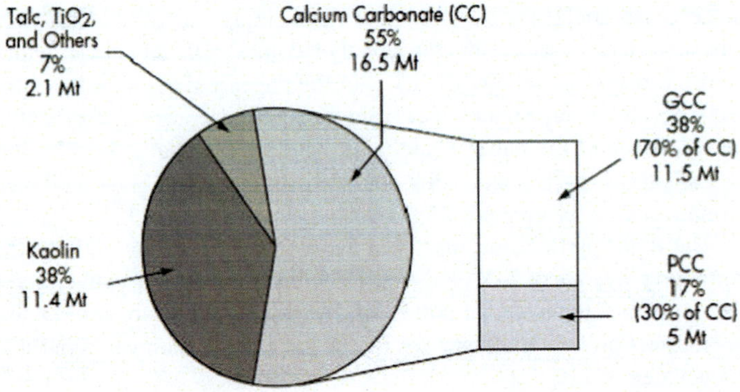

Fig. 3. The most common pigments used worldwide in papermaking (Grönfors 2010:8[25]) (Original source: Harris, R. 2004. Minerals in paper—looking east for growth *Industrial Minerals* 443:52–57.).

A variety of colored and white pigments have been used in order to be given the desired result. In the following table (Table 1) Loureilo et al. [29] composed the white pigments utilized during nineteenth and twentieth century.

Table 1. White pigments of the nineteenth and twentieth centuries (Loureiro LCP et al., 2011:2).

Pigment (and other known names)	**Chemical formula**
Alumina trihydrate	$Al_2O_3.3H_2O$
Barium sulphate (baryta, blanc fixe, heavy spar)	$BaSO_4$
Calcium carbonate, natural from calcite or precipitated from aragonite, (chalk, whiting)	$CaCO_3$
Calcium silicate	Ca_2SiO_4
Calcium sulphate (gypsum, pearl white)	$CaSO_4.2H_2O$
Calcium sulphite	$CaSO_3$
Diatomaceous silica (diatomite)	SiO_2
Hydrated calcium aluminium sulphate (satin white)	$3CaO.Al_2O_3.3CaSO_4.31H_2O$
Hydrated magnesium silicate (talc)	$3M_9O.4SiO_2.H_2O$
Kaolinite (China clay, kaolin)	$Al_2O_3.2SiO_2.2H_2O$
Lead white (hydrocerussite, basic lead carbonate)	$Pb_3(CO_3)_2(OH)_2$
Lithophone (Orr's white, Charlton white)	30% ZnS + 70% $BaSO_4$
Potassium titanate	$K_2O.4TiO_2$
Titanium dioxide (rutile or anatase)	TiO_2
Zinc oxide	ZnO
Zinc sulphide	ZnS
Plastic pigments based on polystyrene (since the 1970s)	

4 Experimental

The sheets were preliminary examined macroscopically and through a digital microscope (Dino-Lite USB Digital Microscope AD4113T-I2V, magnification from 10× t to 50× and 220×). On the basis of this preliminary investigation, the sheets to be further studied by X-ray fluorescence (XRF) analysis were selected. In particular, six sheets were selected to cover all types of paper detected in the artwork: a typewriter sheet (No. 001), two pieces of the most used paper type (No. 281, No. 312) and three papers that are of individual characteristics (No. 130, No. 208, No. 255). Several areas on each sheet of paper were analyzed through XRF, including bare paper (both the *recto* and *verso*), black toner and intermediate hues. In addition, the logotype stamp (an orange cactus) of the manufacturing company of the paper that is observed on the *verso* of the majority of photocopies was examined as well. XRF analysis was conducted using a portable, milli-XRF spectrometer, comprising of a Rh-anode X-ray tube, a Si-PiN diode X-ray detector (XR-100CR, Amptek Inc.) and a multichannel analyzer (MCA 8000A, Amptek Inc.). In order to facilitate comparisons between the analysed spots/sheets, the XRF data (i.e. elemental peak intensities) were normalized through conversion to relative intensities (units: counts per second/microampere – cps/μA) using the following simple equation:

$$X_i = \frac{h_i/t}{I}$$

where: X_i is the relative intensity of element i, h_i is gross area of X_i element's peak (counts), t is spectrum's total acquisition time (seconds) and I is the X-ray generator electric current during acquisition (microampere) [30].

5 Results

X-ray fluorescence analysis allows for the non-destructive elemental identification of materials. Thus this study focuses in the content of fillers and coatings used at the manufacture of the paper. Moreover the toner region of each sheet was measured and compared with the elemental content of the paper matrix in order to obtain information on the composition of the toner. In Table 2 the normalized XRF elemental data are summarized.

Table 2. Summary of the p-XRF data; numbers correspond to normalized elemental peak intensities (cps/μA values).

Sheet	Plain paper (white)									
	Al	Si	P	S	K	Ca	Ti	Ba	Fe	Zn	Other
001r	–	–	0.056	0.023	–	3.533	–	–	0.126	–	Sr
130r	–	–	0.020	0.074	–	0.684	0.915	0.316	tr	0.335	Mn

(continued)

Table 2. (*continued*)

Sheet	Plain paper (white)										
	Al	Si	P	S	K	Ca	Ti	Ba	Fe	Zn	Other
130v	–	–	0.035	0.027	–	0.794	0.172	0.689	–	–	–
208r	–	0.024	0.013	tr	0.084	0.957	–	–	0.163	tr	Rb/Sr
208v	0.013	0.047	0.017	0.015	0.101	1.096	–	–	0.159	–	–
255r	0.008	0.033	0.020	0.015	0.048	1.285	0.058	–	0.198	tr	Mn/Rb/Sr
255v	–	0.018	0.020	0.016	0.054	1.449	0.051	–	0.215	–	–
281r	–	–	–	tr	–	0.619	0.167	–	0.193	7.124	Mn/Sr
281v	–	0.012	tr	0.012	tr	0.578	0.264	–	0.191	6.673	–
312r	–	0.011	–	tr	–	0.410	0.167	–	0.218	7.137	Sr
312v	–	0.017	–	tr	–	0.568	0.263	–	0.205	6.438	–
Sheet	Copy – Black toner										
	Al	Si	P	S	K	Ca	Ti	Ba	Fe	Zn	Other
001r	–	–	0.050	0.020	–	3.261	–	–	0.115	–	Mn
130 gloss	–	–	0.022	0.062	–	0.760	0.802	0.328	tr	0.336	Ag/Br Mn/Sr
130 matte	–	–	0.023	0.057	–	0.714	0.842	0.312	tr	0.338	Ag/Mn
208r	–	0.020	0.019	0.018	0.088	1.188	–	–	0.153	tr	Rb/Sr
208v	0.010	0.039	0.015	0.011	0.093	0.873	–	–	0.158	–	–
255r	–	0.029	0.017	0.017	0.048	1.269	0.062	–	0.218	–	–
281r	–	0.014	0.012	0.021	0.019	0.388	0.149	–	4.516	5.739	Mn/Sr
281v	–	0.017	0.035	tr	–	0.438	0.271	–	2.658	6.479	Mn
312r	–	0.014	–	0.018	–	0.331	0.159	–	4.858	5.436	Mn/Rb/Sr
312v	–	0.019	–	tr	0.024	0.492	0.269	–	2.903	6.222	Mn
Sheet	Copy – grayish areas										
	Al	Si	P	S	K	Ca	Ti	Ba	Fe	Zn	Other
281r	–	–	–	tr	–	0.256	0.162	–	1.113	7.148	Mn
281v	–	0.012	tr	tr	tr	0.475	0.261	–	1.193	6.411	Mn
Sheet	Orange stamp ("cactus"/verso)										
	Al	Si	P	S	K	Ca	Ti	Ba	Fe	Zn	Other
312v	–	0.011	–	0.016	–	0.390	0.259	–	2.670	6.222	Mn

5.1 Fillers and Coatings

Paper sheets numbered No. 281 and No. 312, which correspond to the most common type of paper in the artwork in consideration, are coated with zinc oxide (ZnO). Zinc peak is approximately 15 times higher than the immediate smaller peak which corresponds to calcium (probably calcium carbonate - CaCO3) which has been used as filler. The increase of titanium peak height on the verso by $\geq$60% in comparison to the recto indicates the use of titanium dioxide (TiO2) as filler as well, and highlights the compositional differences between the two paper sides.

Zinc oxide (ZnO) is also detected at sheet coded No. 130 although in this case it is used as an additional element of the titanium (probably TiO_2) coating on the side to be photocopied – *recto* (Fig. 4). Also, this sheet is the only one in which barium (Ba) is found (in a layer under the aforementioned top coat, so barium is better detected on the back). The coexistence of barium (Ba) and sulfur (S) is probably due to the use of the barium sulphate ($BaSO_4$) pigment as a filler. Alternatively sulfur may be due to admixture with calcium to form gypsum ($CaSO_4 \cdot 2H_2O$) or to its simultaneous presence with barium sulphate ($BaSO_4$).

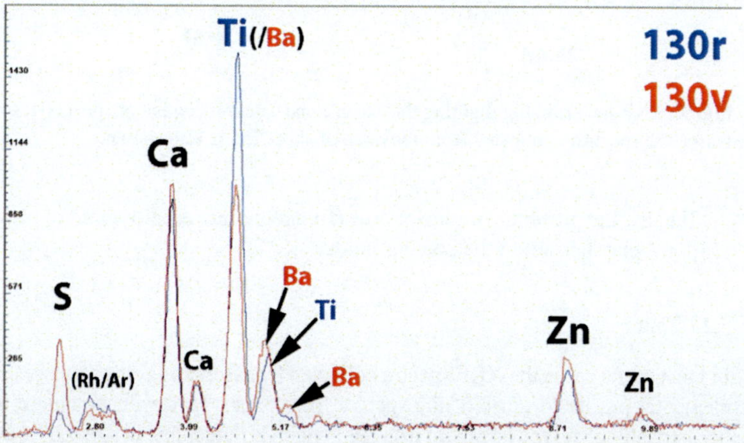

Fig. 4. XRF spectra of recto and verso of sheet No. 130. The recto shows notably less barium yet significant titanium, indicating presence of titanium dioxide coating only on this side.

Sheets No. 208 and No. 255 are not coated. Aluminosilicate content is detected in both of them, while in the second titanium was detected as well. Furthermore the presence of potassium (K) is observed. According to Dabrowski [31], potassium (K) is used in the form of potassium chloride (KCl) for the pH regulation. However, the simultaneous detection of potassium and titanium (Ti) in the case of sheet No. 255 may also pertain to the white pigment potassium titanate ($K_2O.4TiO_2$) (see Table 1). On the other hand, the major elemental peak in both spectra is attributed to calcium (Ca), whose peak is 15 times higher in comparison to that of the other elements, and most probably pertains to the gypsum ($CaSO_4.2H_2O$) as sulfur (S) is also detected.

Last but not least, paper sheet No.001 (the typewriter product) differs from all the other sheets because it shows the highest concentration of calcium (Ca) (Fig. 5) yet no sulfur (S). The exceptionally high calcium (Ca) content is attributed to either the presence of calcium carbonate ($CaCO_3$ used as a filler), or to some other process as calcium is used extensively in paper manufacturing [32].

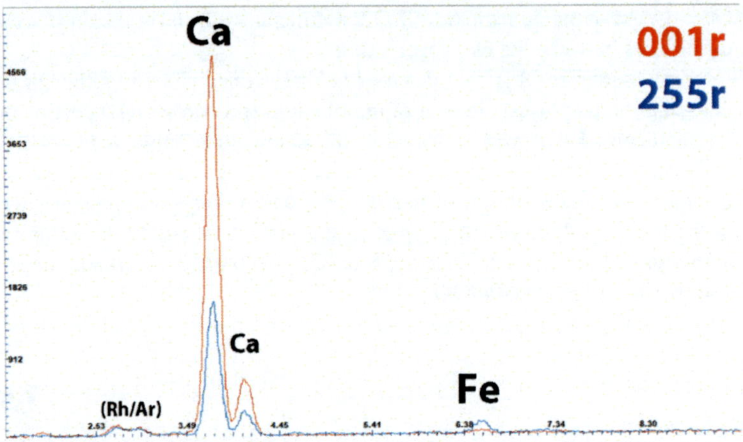

Fig. 5. XRF spectrum highlighting the exceptional calcium content of sheet 001; for comparison reason the spectrum collected from the recto of sheet 255 is also shown.

Finally, the presence of phosphorus (P) may reflect employment of bleaching agent [35] or it may pertain to the coating binder[4] [33].

5.2 Toner

In most of the cases the XRF spectra collected from toner regions so more or less similar elemental profile to those of the spectra from paper. However, notable differences are occasionally observed as regard iron (Fe) which is detectable at sheets No. 001, No. 208, No. 255, No. 281 and No. 312. More specifically while No. 001, No. 208 and No. 255 have the same iron (Fe) peak intensity in both black and plain paper areas, sheets No. 281 and No. 312 show much higher peaks at the toner region (Fig. 6).

4 In addition to pigments, coatings also contain an organic natural or synthetic binder such as starch, casein-phosphoprotein derived from milk, polyvinylacetate latex etc.

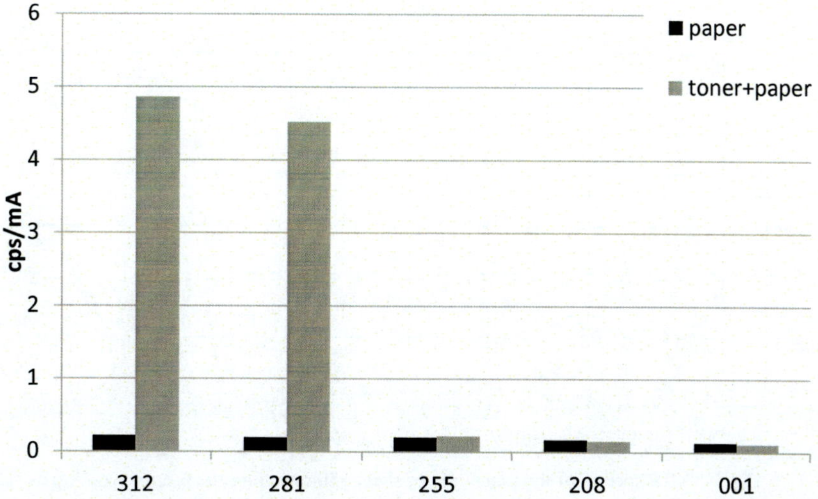

Fig. 6. Comparison of Fe peak intensities for paper and toner regions of sheets coded 312, 218, 255, 208 and 001.

As it is observed in Fig. 6, the differences in iron elemental peak intensities indicate that the iron is a toner component in case of the sheets No. 312 and No. 281. On the contrary, its minor presence at sheets No. 255, No. 208, No. 001 is probably due to water contamination at the manufacturing stages of the paper [34]. This conclusion is drawn because in the last three sheets there is no significant deviation of the concentration of iron in the paper and the toner in order to prove its exclusive presence in the latter. Moreover, high concentration of iron is also observed in the orange stamp on the verso of the sheet, yet this finding is questionable because the XRF analysis might have been affected by the elemental composition of the front side.

During the macroscopic observation of the sheet No. 130, spatial variations were observed in the visual effect of the black areas: in some areas it appears matte while elsewhere it shows a rather glossy appearance. XRF analysis revealed no obvious differences between the glossy and matte areas (Table 2); however, authors were able to identify silver (Ag) along with bromine (Br) in the corresponding spectra (Fig. 7). This finding is rather peculiar and lays in contrast with the carbon-based toners that are commonly used in photocopying processes. In addition, upon observing the photocopy at an angle as well as under the Dino-Lite USB Digital Microscope AD4113T-I2V (Fig. 8), the glossy areas give the feeling of the visual effect silver mirroring presents. Silver mirroring is a degradation process occurring on silver-based photographs shown up as a bluish metallic glint [35]. However, to the best of authors' knowledge there is no literature dealing with the investigation of this phenomenon to photocopying processes and materials, therefore further research towards this direction needs to be done.

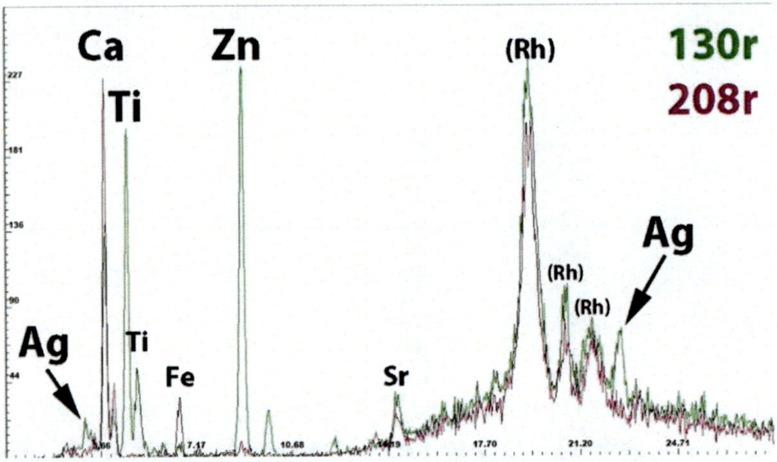

Fig. 7. XRF spectra of the toner area of sheet 130 revealing the presence of Ag, as compared to the corresponding spectra of sheet 208.

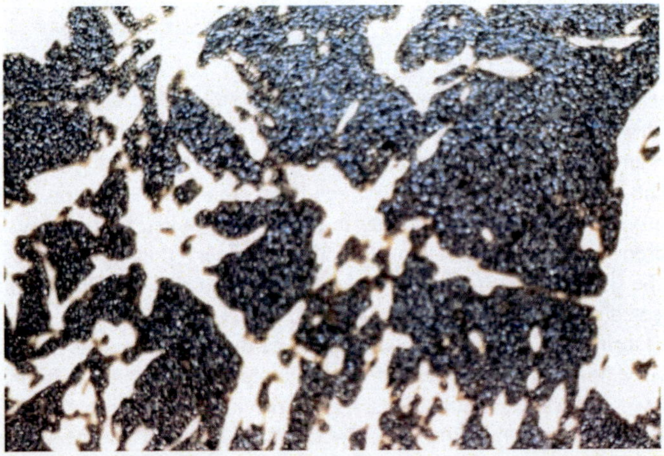

Fig. 8. The mirroring effect of the silvery ink of sheet 130 as it is shown under the digital microscope. (Rena Papaspyrou, "Photocopies directly from the material, 1980–1981", Installation, 305 photocopies of different dimensions (+11 pages with contents), Variable dimensions, Donation of the artist 2014, Entry No. 1007/14, Collection of the National Museum of Contemporary Art, Conservation Department).

6 Discussion

According to the above data the paper sheets are divided into two categories: the typewriter product and the photocopies. Photocopies are subdivided into coated and uncoated paper support and products of direct and indirect process. The sheets No. 218 and No. 312 are coated with zinc oxide (ZnO), No. 130 is coated with titanium dioxide (TiO_2), while the rest of the paper sheets are uncoated. The combination of the zinc oxide (ZnO) coating and the existence of iron (Fe) in the toner indicate that sheets No. 281 and No. 312 have been produced with the direct electrostatic process. On the other hand, the high calcium (Ca) content in paper sheet No.001 is considered reasonable since it is the only filler in the paper matrix.

X-ray fluorescence analysis made it possible to distinguish toners with inorganic content, while for the identification of organic toners it is deemed necessary to be carried out analysis with infrared spectroscopy. For the determination of the physical state of the toner in all the sheets further research is required.

The different visual effect of coated and non-coated papers is that the former present an attractive gloss and sharp contrasts. Obviously, the new copier machine that Athens School of Arts purchased based on indirect electrostatic process and the plain paper could not reach the aesthetic completeness that direct process and coated paper offered to the artist.

7 Conclusions

In the twentieth century, the paper industry burgeoned as technological advances made it easier to make copies and enabled offices and individuals to have photocopy machines on site in contrast to the previous condition that in order to a copy be made anyone should address to a copy center. Although photocopying is a worthless object in most people's minds, some artists have taken advantage of its potential to create remarkable works of art. After all, the artistic world is influenced and always keeps up with technological developments, especially in the context of contemporary art.

Thus, Rena Papaspyrou experimented with the performance of the materials under the light of the photocopier, creating an anthology of the subjects of the urban "invisible" environment/landscape. Her artwork "Photocopies directly from Matter 1980–81", consists of 305 photocopies and 11 content sheets from a typewriter.

In the present study six sheets of this art installation were examined with X-ray fluorescence (XRF) analysis in order to evaluate the different visual effects of paper sheets. These sheets were selected due to their obvious differentiation on paper quality: one typewriter product, two of the paper quality that the multitude of the artwork consists of and three unlike with each other and with the whole. The examination of paper and toner areas made possible to distinguish the coated and uncoated photocopies and consequently to separate them in the direct or indirect photocopying process. The major elements detected are Ca, Ti, Fe, Zn, S, Si and P. A total of three types of toners are identified: a) ferrous copy toner, b) admixture of silver in the toner of one of the photocopies, c) purely organic toner. The differences in the manufacturing technology of the photocopies certainly reflect the replacement of the photocopier with a new one during the creation of the artwork.

Through the results it is concluded that XRF is the most ideal non-destructive method for the identification of modern copies on paper. Nevertheless, for the complete evaluation of the structure of the photocopies, both in the composition of the paper and the toner, a further study of infrared spectroscopy is underway in order to enhance the body of available analytical data and facilitate proper planning of conservation interventions.

Acknowledgements. We thank the National Museum of Contemporary Art in Athens for granting the license to conduct the study and analytical techniques to the artwork under consideration. Our sincerely regards goes to Fotini Alexopoulou, Conservator of Works of Art at Greece's National Museum of Contemporary Art, for her guidance, enthusiasm and support during the whole process. Deepest appreciation is attributed to Elina Kavalieratou, paintings conservator of the National Gallery of Greece and Alexandros Soutzos Museum, for her encouragement to undertake the study, in the context of her collaboration with the National Museum of Contemporary Art. Finally, special thanks are due to Dr Y. Bassiakos and Dr E. Filippaki from INN, NCSR "Demokritos" for providing access to the portable XRF device.

References

1. Mort, J.: Xerography: a study in innovation and economic competitiveness. Phys. Today **47**(4), 32–38 (1994). https://doi.org/10.1063/1.881427
2. Kordic, A.: How did mail art develop into a global art movement? (2016). https://www.widewalls.ch/magazine/mail-art-correspondence-postal. Accessed 13 Apr 2021
3. Walker, J.A.: Copy this! A historical perspective on the use of the photocopier in art. Plagiary Cross-Disc. Stud. Plagiarism Fab. Falsification, 22–24 (2006)
4. Eisen, E.: The work of art in the age of Xerox reproduction (2018). https://thenewinquiry.com/blog/the-work-of-art-in-the-age-of-xerox-reproduction/. Accessed 13 Apr 2021
5. Ginny, L.: 5 Cents a page. Women Artist. News **7**(6), 11–12 (1982)
6. Nochlin, L.: From 1971: why have there been no great women artists? (2015). https://www.artnews.com/art-news/retrospective/why-have-there-been-no-great-women-artists-4201/. Accessed 15 Apr 2021
7. Whitney Museum of American Art, Edwart Meneeley. https://whitney.org/exhibitions/experiments-in-electrostatics?section=2&subsection=1#exhibition-artworks. Accessed 29 Apr 2021
8. Malonee, L.: Travel a familiar world created by a copy machine (2018). https://www.wired.com/story/copy-machine-landscapes-gallery/. Accessed 29 Apr 2021
9. Schizakis, S.: Every month: photocopies straight out of matter, 1980–1981. https://www.emst.gr/en/exhibitions-en/every-month-photocopies-directly-from-matter-1980-1981. Accessed 8 May 2021
10. Papaspyrou, R.: Photocopies directly from matter 1972 and 1980–81. In: Paparounis, M. (ed.) Rena Papaspyrou Photocopies: ΦΩΤΟΤΥΠΙΕΣ ΑΠΕΥΘΕΙΑΣ ΑΠΟ ΤΗΝ ΥΛΗ 1980–1981, Futura, Αθήνα (2011)
11. Totty, R.N.: The examination of photocopy documents. Forensic Sci. Int. **46**, 121–126 (1990)
12. Silbey, J.: How Xerox's intellectual property prevented anyone from copying its copiers, Smithsonian magazine (2019). https://www.smithsonianmag.com/innovation/how-xeroxs-intellectual-property-prevented-anyone-from-copying-copiers-180972536/. Accessed 9 May 2021
13. Gundlach, R.W.: Xerography from the beginning. J. Electrostat. **24**(1), 3–9 (1989). https://doi.org/10.1016/0304-3886(89)90003-X
14. Παυλοπούλου, Δ.: Γραφικές Τέχνες – Η ρεπρογραφία στην Ελλάδα. Σε: Χαρακτική-Γραφικές Τέχνες. Τρίτη έκδοση βελτιωμένη και επαυξημένη, σελ. 164–165, Color Network, Αθήνα (2011) (Greek)
15. Weiss, D.S., Abkowitz, M.: Organic photoconductors. In: Kasap, S., Capper, P. (eds.) Springer Handbook of Electronic and Photonic Materials, pp. 1–1. Springer, Cham (2017). https://doi.org/10.1007/978-3-319-48933-9_37

16. Rai, D.M., et al.: Imaging technology, 2. Copying and nonimpact printing processes. In: Ullmann's Encyclopedia of Industrial Chemistry, Wiley-VCH Verlag GmbH & Co. KGaA, Weinheim (2017). https://doi.org/10.1002/14356007.o13_o08.pub2
17. Vincett, P.S., Sahyun, M.R.V.: Photographic Processes and Materials, In: Meyers, R.A. (ed.) Encyclopedia of Physical Science and Technology, pp.91–132. Third Edition. Academic Press (2003). DOI: https://doi.org/10.1016/B0-12-227410-5/00569-X
18. Hamby, E.S., Gross, E.M.: A control-oriented survey of xerographic systems: basic concepts to new frontiers. In: Proceeding of the 2004 American Control Conference, pp. 2615–2629. IEEE, Boston (2004). https://doi.org/10.23919/ACC.2004.1383860
19. Shein, L.B.: Benjamin Franklin and electrophotography. J. Vac. Sci. Technol. A **25**(4), 1256–1260 (2007). https://doi.org/10.1116/1.2715957
20. Scharfe, M.E., Schmidlin, F.W.: Charged pigment xerography. In: Marton, L. (ed.) Advances in Electronics and Electron Physics, vol. 38, pp.83–146. Academic Press INC., New York (1975). https://doi.org/10.1016/S0065-2539(08)61203-2
21. Holland, N.W.: Photocopy classification and identification. J. Forensic Sci. Soc. **24**, 23–41 (1984)
22. Totty, R.N., Baxendale, D.: Defect marks and the identification of photocopying machines. J. Forensic Sci. Soc. **21**, 25 (1981)
23. Biermann, C.J.: Pulping fundamentals. In: Handbook of Pulping and Papermaking, 2nd edn., pp. 55–100. Elsevier Inc. (1996)
24. Torén, K., Blanc, P.D.: The history of pulp and paper bleaching: respiratory-health effects. Lancet **349**, 1316–1318 (1997). https://doi.org/10.1016/S0140-6736(96)10141-0
25. Grönfors, J.: Use of fillers in paper and paperboard grades. Tampere University of Applied Sciences International Pulp and Paper Technology. Nordkalk Corp., Lic.tech. Lars Grönoors (2010)
26. Ozden, O., Sonmez, S.: Pigments used in the coating of paper and paperboards. In: Arapgirlioglu, H., Atik, A., Hiziroglu, S., Elliott, R.L., Atik, D. (eds.) The Most Recent Studies in Science and Art, vol. 38, pp. 1979–1993. Gece Publicing, Ankara, Turkey (2018)
27. Hubbe, M.A., Robert, A.G.: Fillers for papermaking: a review of their properties, usage practices, and their mechanistic role. BioRes **11**(1), 2886–2963 (2016). https://doi.org/10.15376/biores.11.1.2886-2963
28. Biermann, C.J.: Stock preparation and additives for papermaking. In: Handbook of Pulping and Papermaking, 2nd edn., pp. 190–208. Elsevier Inc. (1996)
29. Loureiro, L.C.P., Sandy, M., Manning S, Seruya, A.I.: European coated papers, 1850–1975: their identification for conservation purposes. In: Bridgland, J. (ed.) Preprints of the ICOM-CC 16th Triennial Conference, paper 0608. ICOM, Lisbon (2011)
30. Mastrotheodoros, G.P., Beltsios, K.G., Bassiakos, Y., Papadopoulou, V.: Two unique Byzantine immured lead-glazed relief ceramic icons and related tiles from the church of St Basil in Arta, Greece: investigation and interpretation of materials and techniques. Archaeol. Anthropol. Sci. **10**, 2059–2074 (2018). https://doi.org/10.1007/s12520-017-0525-2
31. Dabrowski, J.: Fibre loading in papermaking. IPH Paper Hist. **13**(1), 6–11 (2009)
32. Biermann, C.J.: Paper manufacture. In: Handbook of Pulping and Papermaking, 2nd edn., pp. 209–262. Elsevier Inc. (1996)
33. Biermann, C.J.: Pulp bleaching. In: Handbook of Pulping and Papermaking, 2nd edn., pp. 123–136. Elsevier Inc. (1996)

34. Manso, M., Costa, M., Carvalho, M.L.: X-ray fluorescence spectrometry on paper characterization: a case study on XVIII and XIX century documents. Spectrochim. Acta Part B **63**(11), 1320–1323 (2008). https://doi.org/10.1016/j.sab.2008.07.001,p.1321
35. Pietro, G.D.: Examples of using advanced analytical techniques to investigate the degradation of photographic materials. In: Creagh, D., Bradley, D. (eds.) Physical Techniques in the Study of Art, vol. 2, pp. 155–198. Elsevier Science (2007). https://doi.org/10.1016/s1871-1731(07)80006-6

Characterizing the Pigments in Wall Paintings: Comparing Portable and Stationary Multiband Remote Sensing Imaging Systems

Yotam Asscher(✉) and Shai Halevi

Israel Antiquities Authority, Jerusalem, Israel
yotamass@gmail.com, shaih@israntique.org.il

Abstract. Wall paintings are art works with valuable information on ancient technologies, materials and cultural expressions of past civilizations, and assessing wall paintings' state of preservation allows to differentiate between the original materials and degradation products. Imaging techniques such as portable spectral cameras allow studying the materials on a large-scale, and characterizing the pigments and assess their preservation state. A relatively cost effective and easy-to-use method is based on modified digital camera equipped with external filters in the UV-VIS-NIR. The multiband images need to be aligned and normalized to estimate the reflected light's intensity, which makes characterizing pigments difficult, due to noise from uneven illumination, surface degradation, and the physical handling of the external filters. Here, we present a comparison between multiband images from a stationary camera and portable modified digital camera of painted wall fragments, showing the influence of environmental conditions on the quality of the images. The stacked photos of the different filters, representing the reflected intensity at different band widths, were analyzed as multispectral data to study the pigments in the pictorial layer. We combine the data with chemical information from X ray fluorescence (XRF), and high resolution structural information from fiber optic reflectance spectroscopy (FORS), showing portable cameras with external filters could be used as spectral documenting tool for characterizing iron-oxide pigments such as yellow and red ochres on-site.

1 Introduction

Wall paintings are important and valuable works of art that require a significant amount of scientific knowledge concerning their materials for archaeological and conservation investigations. The study should include the assessment of the present state of the pigments, allowing to characterize the materials used in the past and interpret their origin, record past modifications, and define future interventions for conservation strategies (Artioli 2010). Remote sensing was adoption from the world of geology to heritage science for characterize paintings on a large scale in museums, based on reflectance spectral imaging which documents the spectral responses of different materials in high resolution spatial and spectral resolution (Legrand et al. 2014; Dooley et al. 2017; Striova et al. 2020).

A. Osman and A. Moropoulou (Eds.): *Advanced Nondestructive and Structural Techniques for Diagnosis, Redesign and Health Monitoring for the Preservation of Cultural Heritage*, SPM 16, pp. 71–82, 2022.
https://doi.org/10.1007/978-3-031-03795-5_14

The widespread use of digital cameras through the 1990's led to the technological development in the field of spectral documentation in cultural heritage research (Dyer et al. 2013). In the last three decades, digital cameras were used for characterizing materials with quality and improved spatial resolution in the visible region. In addition, the different sensors are sensitive to ultraviolet and infrared radiation, making it a reflectance spectral imaging tool in the UV-VIS-NIR, with the ability to analyze valuable information of materials with cultural significance in different spectral regions (Striova et al. 2020). The two major systems in reflectance spectral imaging are: 1) a camera that documents 2D distribution of the surface through UV-VIS-NIR filters producing multiband images, and 2) imaging spectrometers that produce digital images in a number of contiguous spectral bands (hyperspectral imaging) making a three-dimensional data, termed a spectral cube. The quantification of the reflected light intensity allows for the characterization of different materials based on their response to different wavelength, which is important in the field of remote sensing and material characterization, but the analysis needs to take into account the variations in the reflected intensity due to scattering processes from uneven surfaces, illumination conditions, degradation processes and salts deposits (Slavuj et al. 2013; Zhang et al. 2017).

In the museum, reflectance spectral imaging is combined with additional non-invasive methods for the characterization of pigments, which is of crucial importance for the preservation of valuable paintings, illuminated manuscripts, parchments, and painted statues (Dupuis et al. 2002; Miliani et al. 2007; Artioli 2010; Delaney et al. 2016). One reason for the increase use in non-invasive portable instrumentation is because micro-sampling of intrusive methods is limited in art works of significant importance. Non-invasive methods for detecting and mapping pigments, in two dimensions, are based on non-portable and delicate motorized stages, which are usually limited to the museum environment, combining spectral responses with chemical information. Hyperspectral cameras and macro-X-ray fluorescence spectrometers are placed on these motorized stages, providing maps of reflected responses from different materials and associated chemical information for each pixel (Cucci et al. 2016; Striova et al. 2020).

In the field, characterizing pigments is commonly applied on wall paintings, and is mostly based on multiband imaging techniques, which are cost-efficient and easy-to-use instrumentation compared to high resolution hyperspectral cameras (Blazek et al. 2013). The applications for multiband imaging are mainly in iconography, for identifying hidden decorations, and monitoring changes in color for conservation purposes (Pérez et al. 2013; Chaban et al. 2017). Based on the observations through ultraviolet filters, organics may be mapped allowing to determine the distribution of previous interventions, and using infrared filters, hidden layers and elements may be probed since infrared light has deeper penetration capabilities.

Multiband imaging based on digital modified cameras with external filters has been recently used to characterize pigments (Cosentino 2014; Cosentino 2015). In addition, the distribution of pigments that contain iron oxides, such as hematite and goethite, was mapped using mineral-specific spectral indicators analysis of multiband data on-site. This was achieved by combining the spectral responses of different pigments with complementary non-invasive chemical and structural information from portable XRF spectrometers and fiber optic reflectance spectroscopy (FORS) (Asscher et al. 2019;

Asscher et al. 2021). Modified digital cameras are positioned on a tripod in an angle to external flashes in front of the wall painting, therefore, noises from illumination variability in the field and movements in the positioning of the camera when changing the external band filters need to be taken into account. In this study, we compared a multiband stationary camera that uses fixed positioning and LED lighting, dedicated to produce high resolution spatial information with different bands in the VIS–NIR ranges, to the performance of a portable modified digital camera with 3 external filters in the VIS–NIR. The results were validated to other non-invasive information from FORS and pXRF that provide preliminary information on the composition of the pigments, focusing on iron oxides, discussing the potential of non-invasive multiband imaging to characterize pigments on-site.

2 Materials and Methods

Fragments of wall paintings that were found in secondary deposition ('Ad and Eirikh-Rose 2021), within the foundations of a street in the Roman site of Motza (Israel), were used in this study. Different preservation state and coloring complexity are observed (Fig. 1). The fragments were characterized using non-invasive high resolution reflectance spectrometry (FORS) and portable XRF spectrometer, and the red, pink and yellow colors were imaged using two multiband systems.

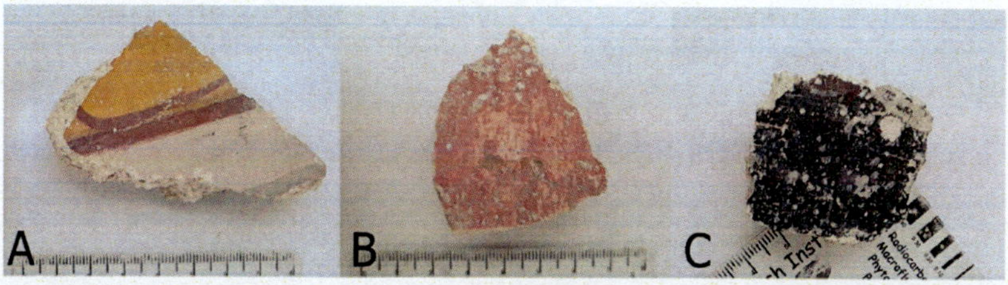

Fig. 1. The painted fragments in this study. (A) a well-preserved decorated fragment b532156 with yellow, red and white colors. (B) a degraded fragment b530399 with pink and green colors. (C) a relatively preserved fragment b532159 with black and red colors.

3 Multispectral Imaging

Portable Camera. In this study, a modified D700 Nikon camera, equipped with a 39-megapixel CMOS full frame sensor (sensitivity of >10% between 300–1000 nm) was used to take multispectral images of the fragments. The camera has been modified, removing the internal UV and IR filters, to collect different images in the UV–VIS-IR range using specific external band-pass filters. A couple of Godox flashes (Godox AD200Pro TTL) including excitations in the UV and NIR ranges was also used to overcome the natural illumination conditions. During the acquisition these flashes are in general positioned at 45° with respect to the measured surface, whereas the camera is placed between the flashlights and pointed orthogonally to the surface (Aach et al. 2009). The multiband images were then calibrated using a Spectralon white that was placed in the photography frame with the fragments during the acquisition. Three external band-pass filters were used in this case to take the images in the Vis–NIR ranges: One for the visible range (about 400–700 nm) and two different filters for the near infrared range (715 nm and 830 nm). The collected photos were normalized, georeferenced and stacked using ENVI v5.0 following Asscher et al. (2021).

Stationary Camera. LED Multispectral Imaging was conducted using a Eureka Vision LED system manufactured by MegaVision, Santa Barbara, CA, USA. The system is equipped with a 39-megapixel Kodak CCD E6 monochrome sensor array, and four Eureka Light LED illumination panels. The camera produces images quantized to a dynamic range of 12 bits per channel. In this method, the spectral component is provided by fixed wavelength light emitting diodes (LEDs) that emit in narrow spectral bands over ranges of wavelengths from the near-ultraviolet to the near-infrared. The LED bandwidth ranges the following wavelengths: 445, 475, 499, 540, 595, 638, 656, 706, 728, 772, 858, and 924 nm. The absolute reflectance is calculated in each wavelength following the correction for illumination gradients, LED power variations and camera noise, using the flat field method, which is based on a Spectralon white (Marengo et al. 2011). The Spectralon covers the full photographic frame, and combined with parallel information from colorchecker targets that are positioned in each image, even values of illumination and coloring profiles (LAB, CMYK, RGB) are produced.

3.1 Fiber Optic Reflectance Spectroscopy

Reflectance spectra of the yellow, pink and black areas in the three fragments were taken using ASD FieldSpec 4 Hi-Res NG fiber optic reflectance spectroradiometer covering 400–2500 nm. Measurements were carried out using a contact probe with a spot size of 10 mm, averaging 30 spectra for statistics and normalized using a Spectralon white.

3.2 X-ray Fluorescence Spectroscopy

X-ray fluorescence (XRF) was carried out on the yellow, pink and black areas in the three fragments. The measurements were carried out using a Bruker 5i Tracer handheld energy dispersive XRF spectrometer. The instrument is equipped with a Rh-anode, miniaturized X-ray tube operating at a maximum voltage of 50 kV and with a Peltier-cooled high-resolution silicon drift detector (SDD). The diameter of the X-ray spot on the sample is about 7–8 mm and accurate positioning on the point to be analyzed is obtained by means of an integrated camera.

4 Results

Spectral reflectance of the painted fragments were based on high resolution FORS measurements, showing the pigments used to color the yellow (b532156), pink (b530399) and black (b532159) areas (Fig. 2). A shift in the position of the iron oxide reflectance peaks is observed in fragment b532156 yellow area, showing the peaks at 600 nm and 770 nm that are associated with yellow ochre (standard from USGS) are influenced by other minerals. The shift in peak position may be associated with a small presence of hematite (Fe2O3) in the yellow ochre. Yellow ochre is composed mainly of clay and sand, with a significant component of the iron oxide limonite ($FeO(OH){\cdot}H_2O$), which may have been oxidized to hematite with time (this transformation is also possible by heating). A similar reflectance spectrum is found in fragment b530399 pinkish area, however, the color is significantly different. This can be attributed to: 1) the bad preservation state of the painted layer, which emphasizes the underlying white plaster, and/or 2) a mixture of yellow and red ochres (with hematite being a dominant component in red ochre), which was used to produce a pinkish color (Elias et al. 2006).

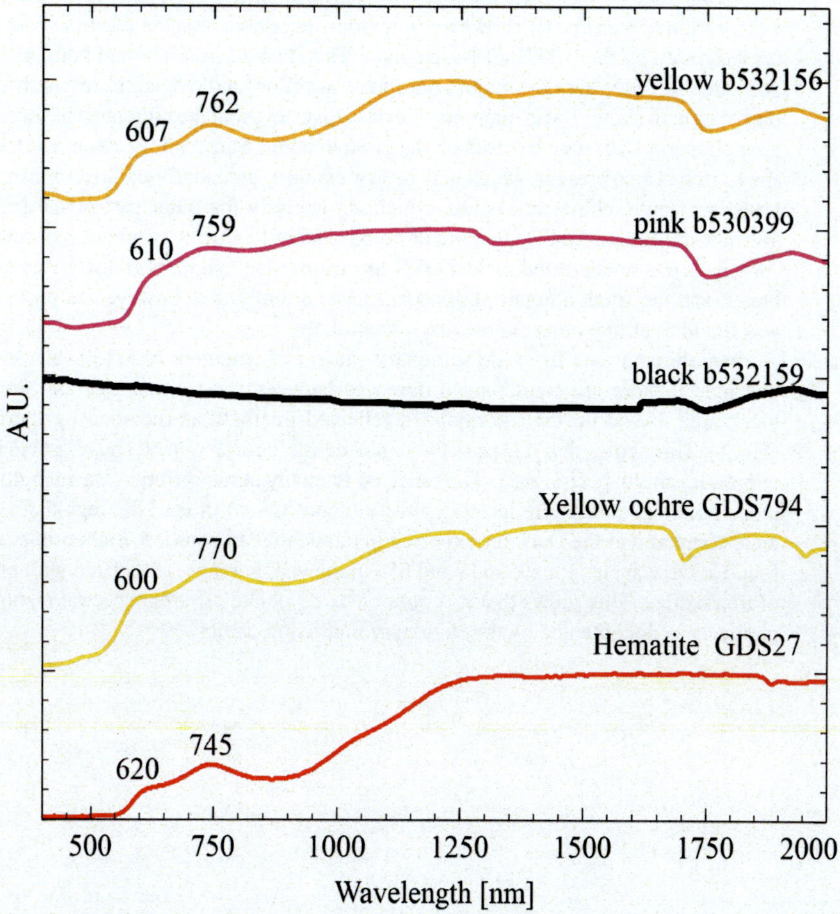

Fig. 2. Reflectance spectra of the yellow area in fragment b532156 (top), pink in fragment b530399 and black b532159, compared to standards from USGS spectral library showing Yellow ochre GDS794 and Hematite GDS27 (bottom).

The green areas in fragment b530399 pinkish area do not influence the spectrum significantly and might be associated with degradation products. The reflectance spectrum of fragment b532159 black area is very poor, which is expected when charred organics are used as pigments (carbon black).

To understand better the reflectance spectra, the same positions were studied for their chemical composition using a portable XRF spectrometer (Fig. 3). The chemical composition of the samples showed calcium (Ca) as a major component, associated with the lime substrate (penetration depth of the X-ray may be as 100 microns, much larger than the painted layer). Minor components were sulfur (S) and potassium (K) associated with salts and degradation products, phosphorus (P) implying there was an organic contamination from the environment (it was found in all the samples), and silicon (Si) and aluminum (Al) which are found in clays that are commonly mixed in lime plaster. The yellow painted area clearly shows a strong iron component, associated with yellow ochre (as suggested by the FORS measurements). The element iron is found both in limonite and hematite, which are the major contributes in yellow and red ochres respectively. The pink fragment shows lower amounts of iron, suggesting that the pigment was mixed prior to application with lime to produce the pinkish color. Surprisingly, the black fragment shows iron is also present, which may be associated to an underlying layer of iron oxide-based pigment, such as red ochre, which is visible in the outer part of the fragment. We rule out the possibility that small amount of red ochre was mixed with the black carbon, as it is not detected in the FORS spectrum. Black pigment is also based on burnt bones, and the small amount of phosphorous can imply that, however, as phosphorous was found in all the other fragments, it is unlikely.

A multiband data from the stationary camera of fragment b532159 (black) shows that indeed there are significant differences between the top red part and the major black area, based on their normalized reflected intensity as recorded by the camera (Fig. 4). This shows that it is possible to normalize, georeferenced and align the images to produce multispectral data. The reflected intensity at the different wavelength shows indeed there is a higher reflected intensity (about twice) in the NIR regions in the red area, compared to the black, as expected in red ochres. In addition, higher reflectance is found at the wavelength 595, 638 and 656 in the visible range, associated with presence of iron oxides. This shows that it is possible to associate different spectral responses to pigments as documented by the stationary multiband camera.

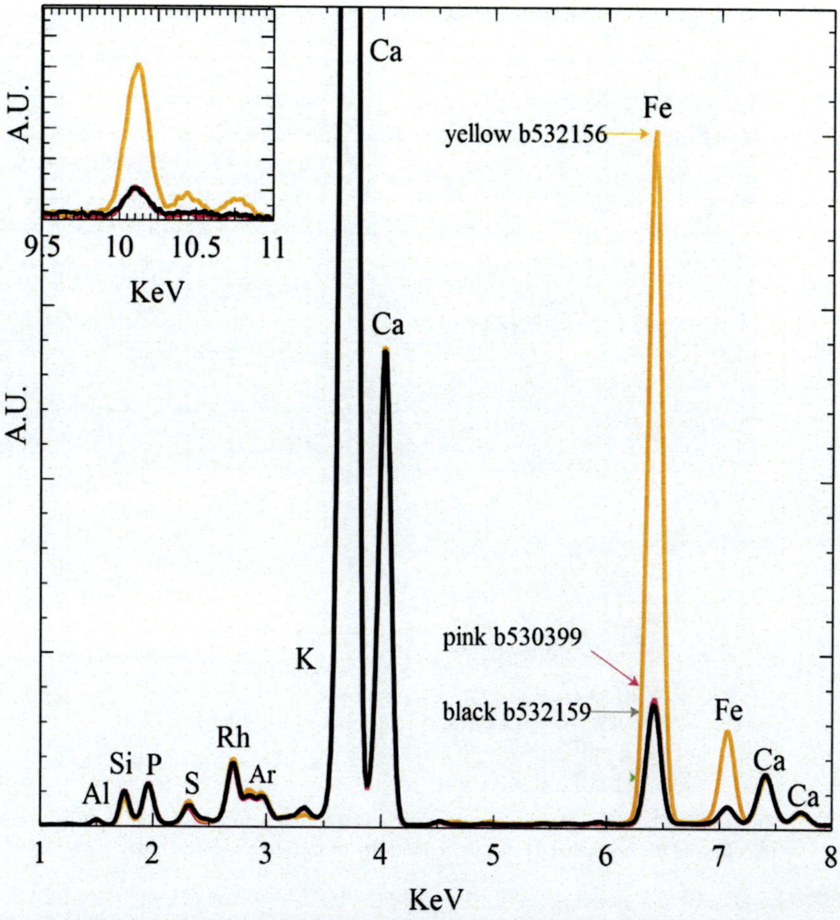

Fig. 3. XRF spectra of the yellow painted area in fragment b532156 (yellow), pink in fragment b530399 (pink) and black in fragment b532159 (black). Inset shows the combination peaks of calcium and iron at 10 keV of the spectra.

A comparison between the stationary and the portable cameras was performed on fragment b532156 (yellow) that was well preserved, showing well defined spectral reflectance (FORS) and relatively high amounts of iron (XRF). The spectral reflectance of the stationary camera shows a clear difference between the red and yellow decorations in the fragment, based on selected areas of interested (Fig. 5A). The normalized reflected intensity shows the yellow area has higher intensity in 550 nm, emphasizing the yellow color, a shoulder at 600 nm and a peak at 772 nm (similar in position to the 762 nm peak found in yellow ochres). The red area is missing the higher reflected intensity at 550 and 600 nm that is found at the yellow and is having a small shoulder at 638 and 656 nm, that might be related to the mixture of red and yellow ochres (Fig. 5B). The portable camera shows very poor spectral resolution, as expected. The comparison between the yellow

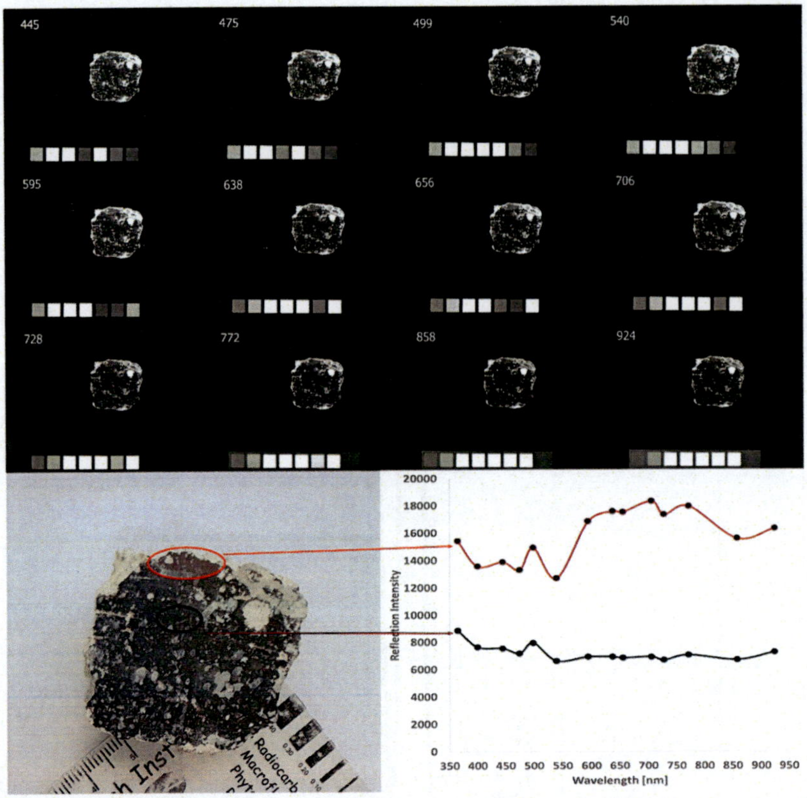

Fig. 4. Multiband images of the stationary camera between 365–924 nm of the fragment b532159 (black), and the normalized reflectance intensity of the red and black areas.

and red areas in fragment b532156 show that higher reflected intensity is observed in the NIR, compared to the VIS (Fig. 5D). In the case of red and yellow colorants, this may be associated with iron oxide pigments. The black fragment shows similar spectral features in the red and black areas of the fragment (b532159). The portable camera does in general show limited possibilities compared with the stationary camera, which would need further verification with XRF and FORS data, to be used to interpret the pigment composition on a larger scale. This could be attributed to the problems in the normalization of the surrounding light, which is based on a single spectralon white, and the need to use strong flashes that could overcome to the surrounding light. The results clearly show how difficult it is to deduce the natural light interference from the surrounding outdoors environment.

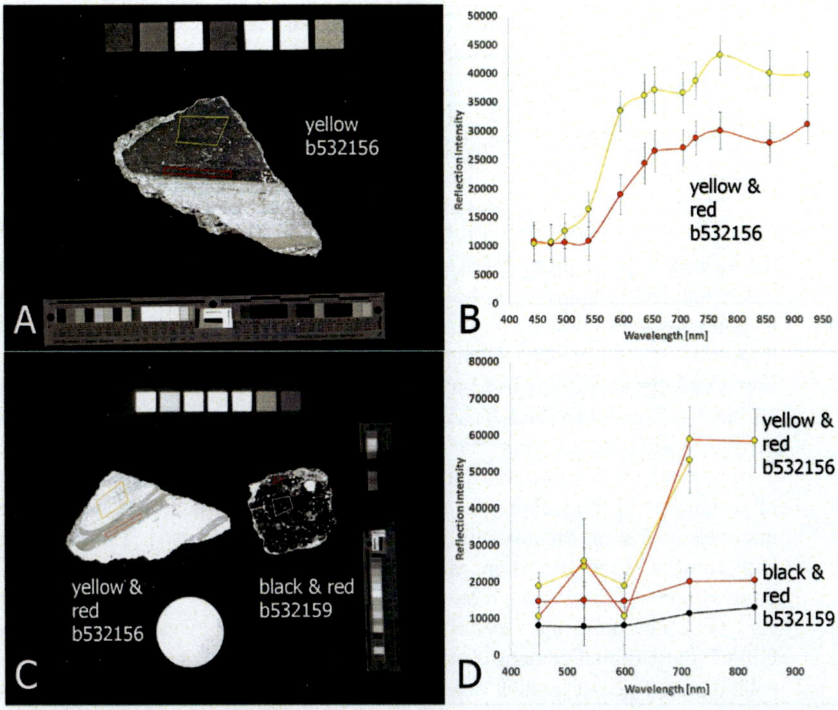

Fig. 5. Images of the stationary and portable camera location of interest, and their associated spectral reflected intensity. (A) The yellow and red areas of interest in fragment b532156, as measured by the stationary camera and analyzed for reflectance intensity that is presented in (B). (C) Fragments b532156 and b532159 as measured by the portable camera, showing the yellow and red locations of interest in b532156, and red and black locations of interest in b532159 that were analyzed for reflectance intensity that is presented in (D).

5 Discussion

Multiband cameras can provide high spatial resolution of cultural remains in 2D, in addition to spectral information in the UV-VIS-NIR regions, showing great potential as non-invasive methods in the toolbox of archaeologists and conservators who come to the field to study pigments in wall paintings. In general, non-invasive methods have an inherent problem of probing only the surface of the wall paintings, due to the limited penetration depth of the UV-VIS-NIR light. Therefore, only the outer painted layers will be characterized, while in depth pigments might be overlooked, in the cases degradation products are not covering the pictorial layer. When combing chemical information (XRF), with high resolution spectral response (FORS), it is possible to interpret better multiband data, which is important on-site since spectral imaging is constrained by the field illumination conditions.

It is important to note that non-invasive methods such as XRF, multiband imaging and FORS open more questions related to conservation, archaeology and contaminations, rather than solving them. They can be used best as prescreening tools on-site, which then allows focusing the efforts on areas of interest and detecting well-preserved painted materials for a better archaeological interpretation and conservation actions. Painted fragments could be taken into the lab, and then fixed positioning and illumination produce better spectral data, with larger number of band-pass filters, which increases the ability to identify pigments. The results obtained in this study show the limitation in the identification of pigments using a portable camera, compared to a stationary one, due to the field illumination conditions. However, the possibilities for the identification of pigments that are based on iron oxides, mainly red and yellow ochres, and black fragments, in well-preserved fragments, is also demonstrated using a portable camera. The portability by itself is less important to fragments of wall paintings, but it is allowing field archaeologists and conservators to aim for low-cost modified digital cameras, instead of high-end expensive equipment in the characterization of pigments.

The application of portable multiband cameras should be more than the identification of pigments and the mapping of their distribution, but to assess preservation state of materials, and limit the possibilities of different materials used. This will allow field archaeologists to include more information that cannot be observed by the naked eye in their interpretations, on a meter-scale, which then can be verified by local spot-size analytical measurements such as FORS and XRF. The documentation on-site using UV-VIS-NIR digital cameras may provide a basis of knowledge on-site, prior to the degradation processes which depend on mass tourism, exposure to oxygen following exposure and other chemical pollutants from the atmosphere.

6 Conclusions

Multiband stationary and portable cameras were used to characterize painted fragments, showing spectral information could provide the identification of iron oxide pigments, such as red and yellow ochres. The spectral images were analyzed for their reflectance spectra in areas of interest, and then combined with chemical information from local measurements using X-ray fluorescence (XRF), and high-resolution spectral reflectance measurements using fiber optic spectrometers (FORS), showing the potential of non-invasive portable techniques to analyze and characterize pigments on-site.

Acknowledgements. We would like to thank Uzi 'Ad and Anna Eirikh-Rose for providing the samples, and Cohava Peterman and Yuval Goren for their advice on the pigments. Moreover, Dr. Ilit Cohen Ofri and Ken Boydstone for working on the stationary multiband system datasets.

References

Aach, T., Brauers, J., Helling, S.: Multispectral image acquisition with flash light sources. J. Imaging Sci. Technol. **53**(3), 31103–31111 (2009)

Ad, U., Eirikh-Rose, A.: Moza, Highway 1. HA-ESI 133 (2021). https://www.hadashot-esi.org.il/Report_Detail_Eng.aspx?id=25953

Artioli, G.: Scientific Methods and Cultural Heritage: An Introduction to the Application of Materials Science to Archaeometry and Conservation Science. Oxford University Press, Oxford (2010)

Asscher, Y., et al.: Combining multispectral images with X-ray fluorescence to quantify the distribution of pigments in the frigidarium of the Sarno Baths, Pompeii. J. Cult. Herit. **40**, 317–323 (2019)

Asscher, Y., et al.: Mineralogical interpretation of multispectral images: the case study of the pigments in the frigidarium of the Sarno Baths, Pompeii. J. Archaeol. Sci. Rep. **35**, 102774 (2021)

Blažek, J., Soukup, J., Zitová, B., Flusser, J., Tichý, T., Hradilová, J.: Low-cost mobile system for multispectral cultural heritage data acquisition. In: 2013 Digital Heritage International Congress (DigitalHeritage), vol. 1, pp. 73–79. IEEE (2013)

Chaban, A., Parisatto, M., Deiana, R., Asscher, Y.: Analysis of on-site multispectral images: a case study of degraded wall paintings in Sarno Baths, Pompeii. In: 2017 Acta Artis Academica Proceedings of the 6th Interdisciplinary ALMA Conference. Augustinian Abbey, Mendel Square, Brno, Czech Republic (2017)

Cosentino, A.: Identification of pigments by multispectral imaging; a flowchart method. Heritage Sci. **2**(1), 1–12 (2014)

Cosentino, A.: Panoramic, macro and micro multispectral imaging: an affordable system for mapping pigments on artworks. J. Conserv. Mus. Stud. **13**(1) (2015)

Cucci, C., Delaney, J.K., Picollo, M.: Reflectance hyperspectral imaging for investigation of works of art: old master paintings and illuminated manuscripts. Acc. Chem. Res. **49**(10), 2070–2079 (2016)

Delaney, J.K., Thoury, M., Zeibel, J.G., Ricciardi, P., Morales, K.M., Dooley, K.A.: Visible and infrared imaging spectroscopy of paintings and improved reflectography. Heritage Sci. **4**(1), 1–10 (2016)

Dooley, K.A., Coddington, J., Krueger, J., Conover, D.M., Loew, M., Delaney, J.K.: Standoff chemical imaging finds evidence for Jackson Pollock's selective use of alkyd and oil binding media in a famous 'drip'painting. Anal. Methods **9**(1), 28–37 (2017)

Dupuis, G., Elias, M., Simonot, L.: Pigment identification by fiber-optics diffuse reflectance spectroscopy. Appl. Spectrosc. **56**(10), 1329–1336 (2002)

Dyer, J., Verri, G., Cupitt, J.: Multispectral imaging in reflectance and photo-induced Luminscence modes: a user manual. British Museum (2013)

Elias, M., Chartier, C., Prévot, G., Garay, H., Vignaud, C.: The colour of ochres explained by their composition. Mater. Sci. Eng. B **127**(1), 70–80 (2006)

Legrand, S., et al.: Examination of historical paintings by state-of-the-art hyperspectral imaging methods: from scanning infra-red spectroscopy to computed X-ray laminography. Heritage Sci. **2**(1), 1–11 (2014)

Marengo, E., et al.: Technique based on LED multispectral imaging and multivariate analysis for monitoring the conservation state of the dead sea scrolls. Anal. Chem. **83**(17), 6609–6618 (2011)

Miliani, C., Rosi, F., Borgia, I., Benedetti, P., Brunetti, B.G., Sgamellotti, A.: Fiber-optic Fourier transform mid-infrared reflectance spectroscopy: a suitable technique for in situ studies of mural paintings. Appl. Spectrosc. **61**(3), 293–299 (2007)

Pérez, M.C., et al.: Ariadne's house (Pompeii, Italy) wall paintings: a multidisciplinary study of its present state focused on a future restoration and preventive conservation. Materiales de Construcción **63**(311), 449–467 (2013)

Slavuj, R., Green, P.: To develop a method of estimating spectral reflectance from camera RGB values. In: 2013 Colour and Visual Computing Symposium (CVCS), pp. 1–6. IEEE, September 2013

Striova, J., Dal Fovo, A., Fontana, R.: Reflectance imaging spectroscopy in heritage science. La Rivista del Nuovo Cimento, pp.1–52 (2020)

Zhang, X., Wang, Q., Li, J., Zhou, X., Yang, Y., Xu, H.: Estimating spectral reflectance from camera responses based on CIE XYZ tristimulus values under multi-illuminants. Color. Res. Appl. **42**(1), 68–77 (2017)

Nondestructive Evaluation of the Pathology of the Katholikon of the Monastery of Panagia Varnakova with Ground Penetrating Radar

Vasileios Keramidas, Kyriakos Lampropoulos(✉), George Bletsas-Yfantis, Elisavet Tsilimantou, Charalampos Mouzakis, and Antonia Moropoulou

National Technical University of Athens, Iroon Polytechniou 9, 15780 Zografou, Greece
klabrop@central.ntua.gr

Abstract. The Monastery of Panagia Varnakova, dedicated to the Assumption of Virgin Mary, is a historic Byzantine monastery, at Efpalio Municipality, east of Nafpaktos, Greece. The Katholikon has sustained damage from the Aigio earthquake in 1995 and remains closed since 2010. This work presents the nondestructive evaluation of the pathology of the monument through the complementary utilization of geometric documentation data, mapping of macroscopic cracks, nondestructive prospection of the structure of Katholikon with ground penetrating radar, finite element modeling and assessment of the dynamic behavior of the Katholikon under simulated earthquakes. The results from the ground penetrating radar coincide significantly with model predictions, demonstrating the complementarity of nondestructive testing with modeling. This information is crucial for the design of appropriate restoration and strengthening measures that will address the critical areas of the building.

Keywords: Nondestructive testing · Ground penetrating radar · Finite element modelling · Varnakova Monastery

1 Introduction

The Monastery of Panagia Varnakova, dedicated to the Assumption of Virgin Mary, is a historic Byzantine monastery at Efpalio Municipality east of Nafpaktos, Greece. The monastery was built in 1077 by St. Arsenios Varnakovitis, and it was financially assisted by the Komnenoi Byzantine emperors, due to its miraculous icon of Virgin Mary. Many Komnenoi Emperors were buried within the Katholikon (main church) of the Monastery, indicative of its religious significance. The Katholikon is a basilica with a cupola, consisting of the nave and aisles, decorated with wonderful opus sectile marble floor, an eso-narthex, built in 1151, and an exo-narthex, built in 1229. The Katholikon was set on fire in 1700, restored in 1805 and partially destroyed in 1826 by the Turks, but renovated with financial aid from Ioannis Kapodistrias, first Governor of Greece, in 1831. The Katholikon sustained damage from the Aigio earthquake in 1995 and remains closed for security reasons since 2010. Metal buttresses have been installed to provide

A. Osman and A. Moropoulou (Eds.): *Advanced Nondestructive and Structural Techniques for Diagnosis, Redesign and Health Monitoring for the Preservation of Cultural Heritage*, SPM 16, pp. 83–95, 2022.
https://doi.org/10.1007/978-3-031-03795-5_16

stability and strengthen the building. On 29th Jan. 2017 a fire erupted that destroyed many buildings around the Katholikon, mainly the monastic cells, and the archives of the Monastery were lost. On 14th Jun. 2020 the Monastery was again damaged by a new fire, which completely destroyed the new church that was built after the closure of the Katholikon, and unfortunately the historic miraculous icon of Panagia Varnakova as well as other religious relics.

Recently, the National Technical University of Athens (NTUA) has conducted a diagnostic study on the Katholikon of the Varnakova Monastery, to assess its state of preservation, reveal the construction stages, identify areas that need reinforcement interventions and design and propose appropriate restoration works to be undertaken to the Katholikon. The restoration works on the Katholikon were inaugurated on 17.07.2021 and are expected to last a year. This work presents the nondestructive evaluation of the state of preservation of the structural system of the Katholikon of the Monastery of Panagia Varnakova, through the correlation of pathology data, ground penetrating radar prospections and finite element modeling of the dynamic behavior of the Katholikon structure.

2 Pathology

The monastery is built on a hill that rises about 760 m which is a flysch formation (alternating layers of sandstone and clay) that usually consists of folded layers with cracks, ridges and a variety of slopes that may pose risks to the foundation of the Katholikon. However, no significant problems in superstructures due to subsidence are observed macroscopically.

However, extensive cracks are observed macroscopically throughout the cell of the Katholikon (Fig. 1). At the north masonry of the Katholikon, a cluster of cracks on both sides of the junction of the masonry with the interior transverse wall that seperates the nave with the exo-narthex is documented. Many other cracks originate at the openings (doors and windows) of the Katholikon.

The west masonry of the Katholikon is also extensively cracked especially at the area above the main entrance. At the interior, a large diagonal crack is observed on the north wall of the nave, as well as various other cracks in both north and south walls of the nave and exo-narthex and at the niche of the Ieron. These cracks are accompanied with deviations from verticality for the north and south walls. Specifically, the north and south walls have bent outwards, about 2.5 cm each, along the entire length of the Katholikon, while the respective columns of the nave seem to have shifted outwards approximately 4–4.5 cm each at the height of the capitals. These are indications of an out-of-plane bending of the two walls. In the East–West direction, the north wall of the nave does not show any significant deviation, whereas at the south side of the Katholikon, a deviation of approximately 4 cm is observed at the height of the capitals.

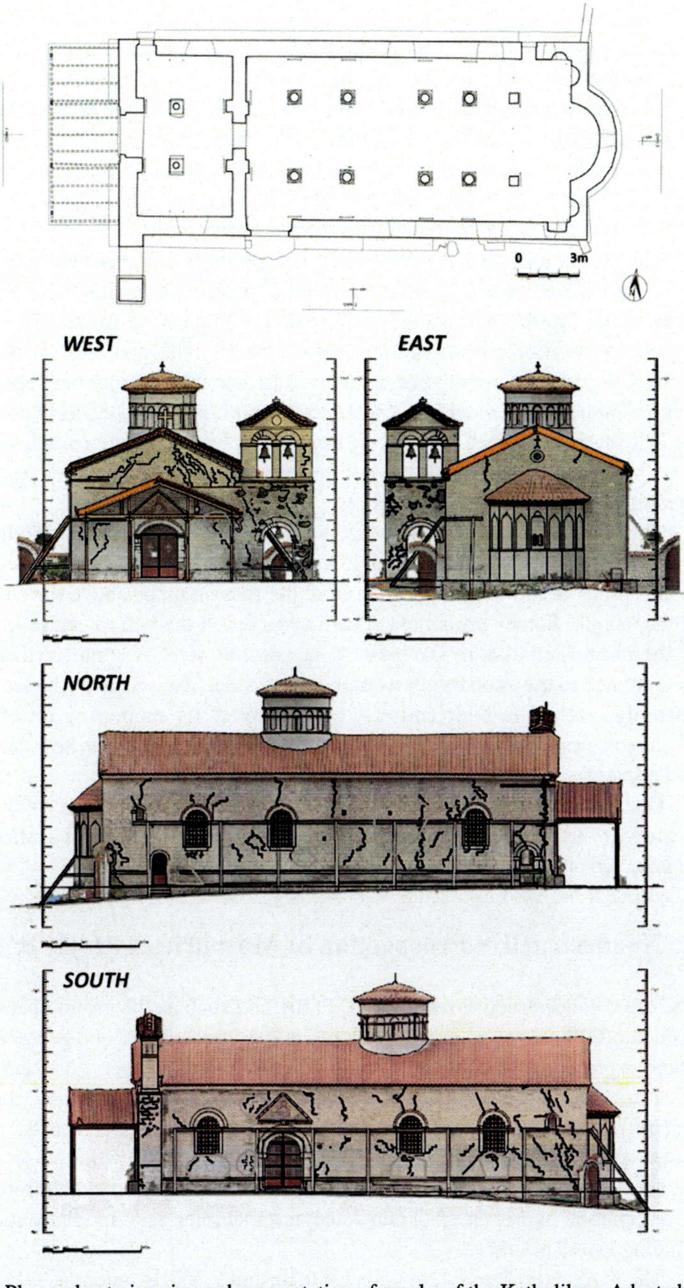

Fig. 1. Plan and exterior views, documentation of cracks of the Katholikon. Adapted from [1]

The two square pillars of the Ieron do not show noticeable deformations, whereas the columns of the main nave appear to deform in a similar pattern as that of the arches that support the dome. Specifically, the four cylindrical columns between the dome and the Ieron having turned about 2.5 cm to the east, whereas the four cylindrical columns between the dome and the exo-narthex have turned about 2.5 cm to the west. In the exonarthex, the western wall at the height of the capitals of the pillars of the main temple seem to have turned to the west between 4.5–5 cm, except on its south side in contact with the bell tower, where the deviations are insignificant.

The observation that the west wall (i.e. the main entrance) has shifted to the west more than the columns of the main temple, possibly explains the significant vertical crack of the north wall at the junction of the main nave-exonarthex, as well as the presence of reinforcements at the corner between north and west. It is interesting that such reinforcements are not present at the corresponding corner between west and south walls, although that corner is in contact (and thus bounded in terms of deformation) with the bell-tower. The north pillar of the exonarthex has turned approximately 3.5 cm to the west, however, the south column has turned to the east, a situation that may exert shear stresses in the vault of the exonarthex.

The north column of the nave that is closest to the Ieron shows significant horizontal cracks in its base as well as a notable detachment of material. Cracks are also observed in the aprons of the windows and around the mosaic niche above the southern entrance of the temple. Some vertical cracks are observed at the bell tower, in-level at the walls of the Ieron (and also in the western façade), as well as some vertical cracks above the entrance to the exonarthex from the north side. The vaults of the katholikon appear intensely cracked in their entirety, with many of its supporting arches on the verge of collapse, especially taking into account that several of their keys have shattered or collapsed.

Due to its state of preservation and the macroscopically observed cracks and in order to prevent further deformations and potential collapsing, a metal scaffolding has been constructed around the Katholikon.

3 Nondestructive Prospection of Masonries with GPR

The use a nondestructive testing (NDT) method such as the ground penetration radar – georadar (GPR), in combination with historical, architectural and geometric documentation, can provide invaluable information for the structural layers and state of preservation of cultural heritage assets and infrastructures [2–4]. GPR uses an electromagnetic pulse that diffuses and travels through the prospected structure and detects the reflected pulses on the examination surface. The diffusion and reflection of the pulse depends on the electromagnetic properties of the building materials that it travels through. Due to this differentiation in properties, it can detect internal interfaces and discontinuities, such as building layers or cracks.

The MALA ProEx system with 1.6 GHz and 2.3 GHz antennae was utilized for the GPR prospection of the masonries in order to reveal their internal layers (masonry-type) as well as to verify whether the documented cracks were extended throughout the width of the masonries, thus, posing significant structural risks. 129 radargrams were filtered and analyzed with the MALA RADExplorer 1.41 software. Due to the lack of access scaffolding at the exterior and interior areas of the Katholikon, at the examination period, GPR scans were conducted at a height accessible as possible.

Figure 2 shows typical segments of GPR radargrams conducted at the north exterior surfaces of the Katholikon. These two radargrams are overlaid on the plan of the Katholikon (black outline), and regard the eastern section of the north masonry (see inset D – red outline zone). These scans are typical of other scans in this area.

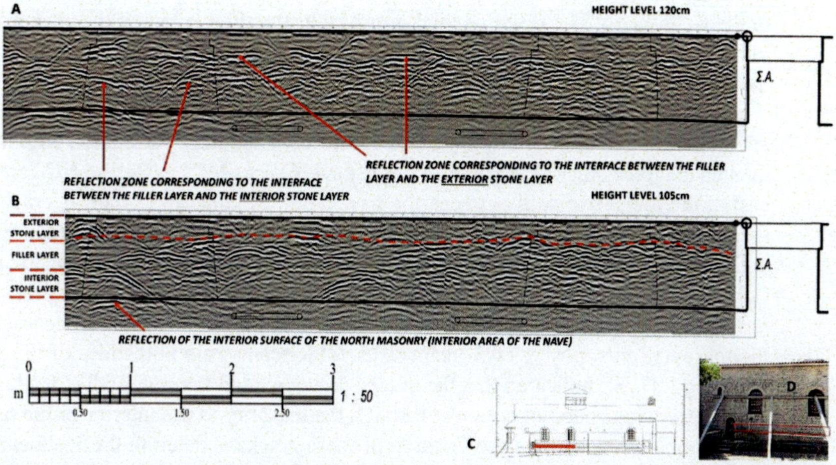

Fig. 2. Segments of GPR radargrams overlaid on the plan of the Katholikon (black outline), in the eastern section of the north masonry (see inset D – red outline zone). A. GPR scan at height level 120 cm, with indication of reflections corresponding to the interfaces observed. B. Height level – 105 cm, indicative curve corresponding to the interpreted location within the width of the masonry of the interface between the exterior stone layer and the filler layer. C. Location (red line) of the above sections of the scans.

According to the analysis of this masonry, this is a three-leaf type masonry. Specifically, as shown in Fig. 2A two systematic zones of reflections can be identified. The first zone of reflections, i.e. the one closest to the exterior surface, corresponds to the interface between the exterior stone layer and the filler layer (i.e. the middle layer). Figure 2B depicts an indicative curve (dashed red curve) corresponding to the interpreted location within the width of the masonry of the interface between the exterior stone layer and the filler layer. After the application of a velocity model, the GPR scans vertical time-scale was converted to depth-scale, enabling overlay on the Katholikon plan and aid identification of targets. Thus, the first interface, which is located at a depth of about 25 cm from the outer surface of the building, corresponds to the interface between the outer stone structure (approximately 25 cm thick) and the inner filling layer.

The second interface, by analogy, which is located at a depth of about 25 cm from the inner surface of the building, corresponds to the interface between the inner stonework (approximately 25 cm thick) and the inner filling layer. The intermediate area corresponds to filler. Therefore, the masonry is three-layered, with the thickness of the two stone structures (external and internal) about 25 cm each, while the rest of the thickness of the masonry consists of filler material (probably stones of irregular shape and size and mortar). It was not possible to detect and confirm, due to multiple internal reflections, whether there are interface stones between the two layers of masonry. As expected, the same layer thicknesses were detected for the case of the scans over the south masonry of the Katholikon. The revealed thickness of the structural layers of the masonries were introduced in the FEM model of the Katholikon, whereas materials properties were measured as part of the diagnostic study [1]

Figure 3 describes the GPR assessment of the state of structural cracks in the western segment of the north masonry of the Katholikon and their correlation with the documented diagonal crack at the interior of the nave. Cracks NC5, NC4 and NC3 are located east of the transverse wall between the nave and the exonarthex whereas crack NC6 is locate in the exonarthex area (Fig. 3A, B). There is observed a diagonal crack (Fig. 3C), at the interior side of the masonry, at a corresponding position opposite (i.e. at the interior) of cracks of NC3, NC4 and NC5. GPR was utilized to assess whether these cracks are connected, in which case the cracks would extend throughout the whole thickness of the masonry, thus, posing significant risks. GPR scans from both sides of the masonry area (Fig. 3D, E), indicated that the cracks do not extend through both stone layers, i.e. the exterior cracks do not continue through the masonry to the interior of the nave.

It should be noted that the creation of these cracks is dated to the 1995 earthquake and is correlated with the influence between the transverse wall between the nave and the exonarthex and the north masonry of the Katholikon.

The "mirror" area, at the south side of the Katholikon presents a different pathology. Specifically, in the area corresponding to NC5 and NC4, a large entrance door is located, with various cracks around it (Fig. 1). A series of horizontal and vertical GPR scans were conducted at the south side of the Katholikon, in order to study the internal layers of the masonry and its state of preservation.

Although the thickness of the stone and filler layers were revealed to be equivalent to those of the north side of the Katholikon, the reflection patterns of the GPR pulse around the south entrance door are interpreted to be corresponding to a detachment of the exterior layer. Specifically, Fig. 4 shows a typical GPR scan conducted at a height level of 120 cm, west of the south entrance door, as depicted in the inset photo.

The reflection zones corresponding to the interfaces between the filler layer and the exterior and interior stone layers are again observed, as in the case of the north side (see Fig. 2). However, in the examined areas of the south masonry (Fig. 4 blue rectangular outline) the areas around of the door (Fig. 4 yellow colored) exhibited more intense reflection zones between the filler and the exterior stone layers, as compared to the remaining masonry areas. This was interpreted as an indication of detachment of the exterior layer from the filler layer.

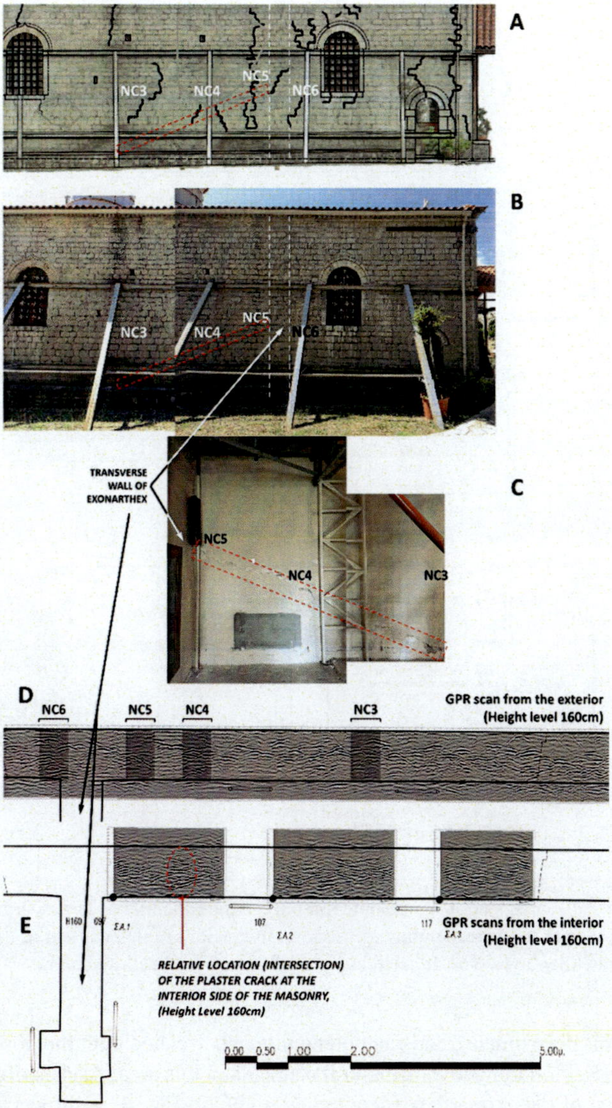

Fig. 3. Assessment of the state of structural cracks in the western segment of the north masonry of the Katholikon and their correlation with the documented diagonal crack at the interior of the nave. A. Location of the cracks on the north view of the Katholikon. B and C. Photos of the actual area. D. Segment of GPR scan from the exterior surface of the masonry. E. GPR scans from the interior of the masonry. D & E overlaid on the plan of Katholikon.

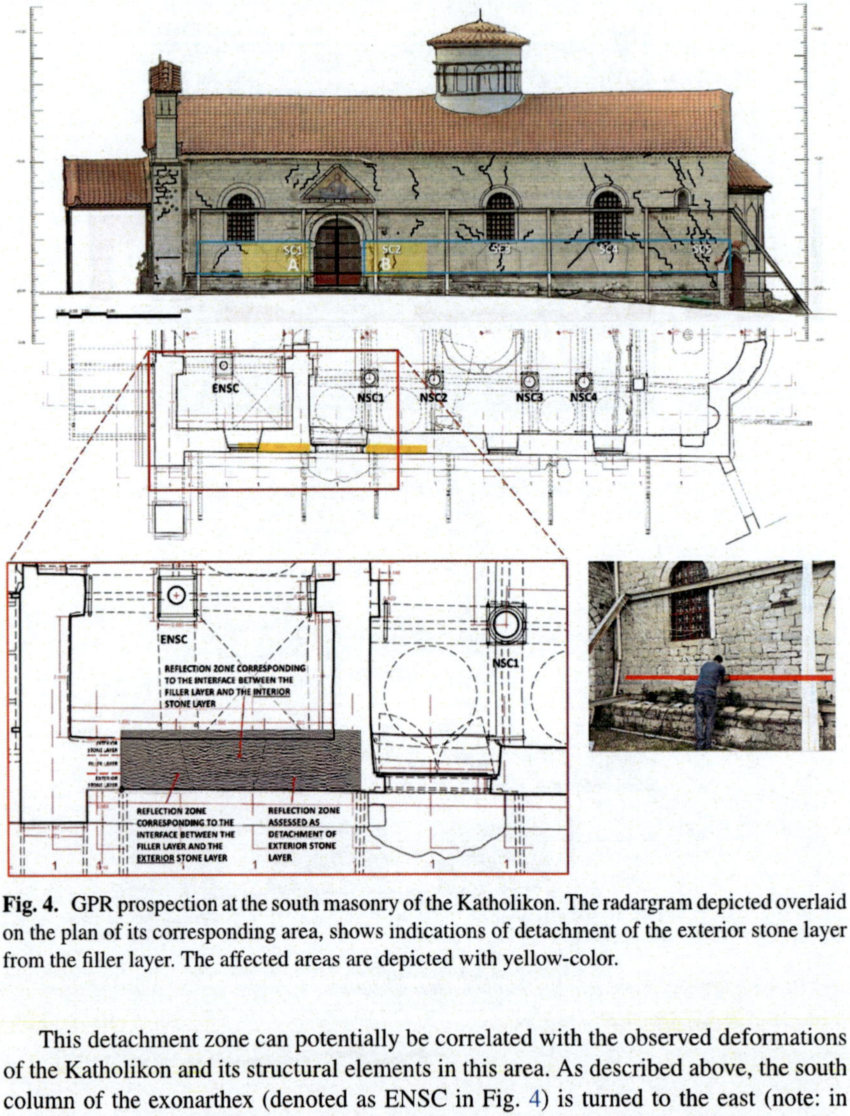

Fig. 4. GPR prospection at the south masonry of the Katholikon. The radargram depicted overlaid on the plan of its corresponding area, shows indications of detachment of the exterior stone layer from the filler layer. The affected areas are depicted with yellow-color.

This detachment zone can potentially be correlated with the observed deformations of the Katholikon and its structural elements in this area. As described above, the south column of the exonarthex (denoted as ENSC in Fig. 4) is turned to the east (note: in contrast to the corresponding north column which turns to the west). In addition, the four cylindrical columns between the dome and the exonarthex have turned about 2.5 cm to the west (note: the south pair is denoted as NSC1 and NSC2 in in Fig. 4). Furthermore, as observed macroscopically and described above the north and south walls have bent outwards, about 2.5 cm each, along the entire length of the Katholikon. If we take into account that in the corresponding area at the north side (Fig. 3) there does not exist any significant opening as in the case of the south entrance, the distribution of shear stresses in this area may have resulted in the form of detachment of the exterior stone layers, up to the genesis of the door arch. The affected areas at both sides of the south entrance door are indicated with yellow color.

The remainder masonry reveals similar behavior as that of the north masonry, i.e. the observed cracks (e.g. SC3, SC4, SC5) do not appear to be extending through the whole width of the masonry, only limited to the 25 cm width of the exterior stone layer.

4 FEM and Dynamic Behavior of the Katholikon

The 3D-model (Fig. 5) of the Katholikon was based on the geometrical and architectural documentation of the structure with the aid of 3D laser scanning, photogrammetric techniques, for the creation of the point cloud. The 3D models of the structure were created with AutoCAD, Revit and Inventor for the smoothing of areas where their geometry was vague. Due to the lack of detailed information about the roof, two models were created, one with the Katholikon roof and one without, i.e. only the bearing the structure) for the numerical analysis. The analysis [5] was implemented with the finite element software ABAQUS 6.14. The model included the Katholikon, the Bell tower and their foundations.

The Katholikon, the bell tower and the foundation consist of load-bearing masonries. The discretization of the model was achieved using the free-mesh technique. The complete cell was discretized with 3D continuous solid elements. The maximum FE dimensions per segment of the cell was adjusted such that at least two FEs described the thickness of the masonry, whereas the minimum FE dimension corresponded to the typical dimensions of the stones used in the masonry construction, including the joint mortars.

The thickness of the structural layers of the masonries of the model are based on the diagnostic study [1] regarding the type of materials and their mechanical properties [6]. The resultant FEM (Fig. 5) consists of 285.998 nodes, 174.713 finite elements and 875.994 degrees of freedom.

The static loads on the bearing structure are imposed from the self-weight of the various building materials. They are calculated from the geometrical dimensions of the components and the specific weight of each building material.

In addition to the static loads from its constitutive materials, the bearing structure of the Katholikon was assessed for the seismic forces that may threaten its structural integrity. The seismic forces were based on the historical seismicity of the Fokida region, according to the current provisions of Eurocode 8 [7, 8]. According to EC-8 this is a religious building with visitors attendance, and it is categorized as CC3-a. The examined limit states were Near Collapse (NC) and Significant Damage (SD). According to the hazard map of the region, the Monastery is located at a seismic risk zone Z2. In order to calculate the acceleration spectrum, the provisions of the current Eurocode 8, EC-8 [8] are applied for Type III design earthquake spectrum with a reference peak ground acceleration of $a_{gR} = 0.24$ g, ground type B, and the maximum value for the importance factor $\gamma_{III} = 1.40$.

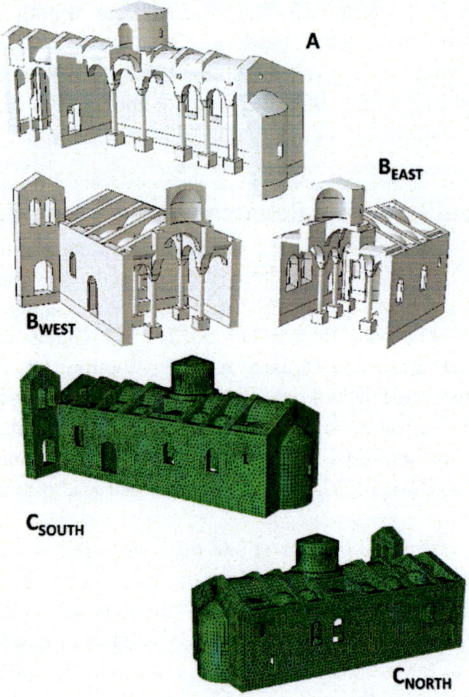

Fig. 5. Longitudinal (A) and Transverse (B) sections of the 3D model of the Katholikon and the Finite Element Model (C) [5]

The model of the Katholikon was subjected to the following types of analysis: a) analysis for static loads; b) modal analysis, MA, to determine the eigenmodes and eigenvectors; and c) modal response spectrum analysis, MRSA, applying the design spectrum. The calculations were based on the consideration that the Katholikon structure is monolithically supported by a homogeneous elastic half-space.

The first significant modes of the Katholikon along the horizontal direction x-x (North–South) and y-y (East–West) are depicted in the Fig. 6. The model was initially considered without damage. Then it was subjected to a simulated earthquake and the eigenperiods resulting from modal analysis were compared with the eigenperiods that have been measured in-situ with modal testing [9], and the model was adjusted accordingly. Specifically, from the analysis of the in-situ measuements the fundamental eigenfrequency at direction north–south is found to have a value of 2,93 Hz, whereas the fundamental eigenfrequency at the direction East–West has a value of 4,88 Hz. The difference in these eigenfrequencies is attributed to the fact that at the direction East–West there exist the north and south walls which function through their in-plane direction. Although the range of microvibrations in which measurements were conducted is rather small, the difference in the measured eigenmodes indicates the lack of cooperation of the parts of the building after the extensive cracking of the monument.

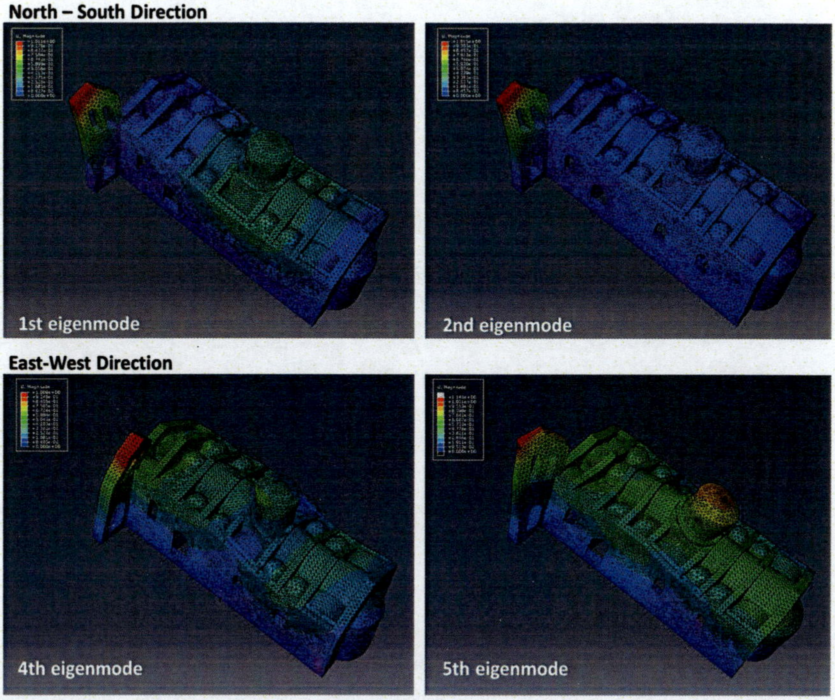

Fig. 6. Dominant eigenmodes of the Katholikon: Top: first and second dominant modes along x-x direction (N-S); Lower: fourth and fifth dominant mode along y-y direction (E-W).

Figure 7 shows the contours of maximum principal stresses for the Katholikon structure. The areas of principal stress strength exceedance are depicted with gray color. Based on the initial assumptions and actual in-situ measurements [6] regarding the mechanical characteristics of the Katholikon masonries the analysis yields maximum stresses at various locations on the cell of the Katholikon, on its vaulting system and at the bell tower.

The above analysis coincides well with the pathology of the monument (Fig. 1), especially regarding the areas discussed above, such as the intersection of the internal wall separating the nave and the exonarthex with the north masonry (Fig. 3) and the area around the south entrance (Fig. 4).

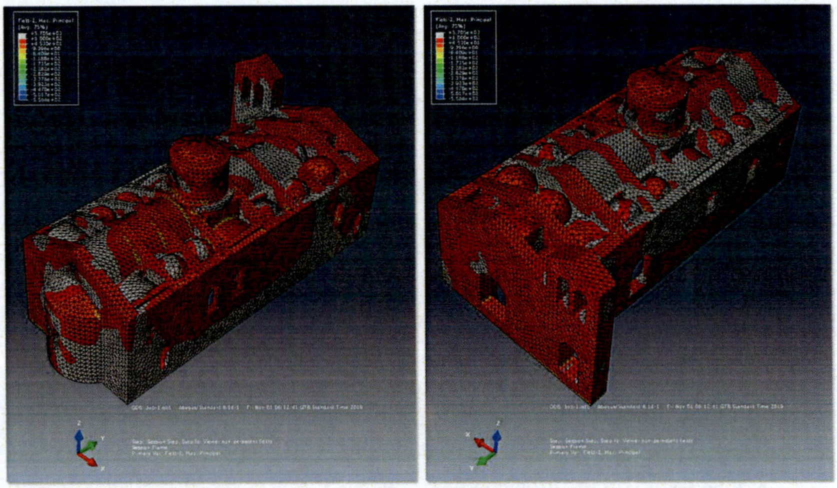

Fig. 7. Maximum principal stresses under seismic loading (Fiuli-type, t = 4,8 s) for the Katholikon of Varnakova Monastery. Gray color indicates areas of principal stress exceedance.

5 Conclusions

This work demonstrates the complementarity of the geometric documentation, the documentation of pathology, the nondestructive prospection of the structure, the finite-element analysis and the assessment of the seismic behavior of the monument, in providing a more representative "picture" of the actual dynamic behavior of this monument that has sustained significant damage. This information is crucial for the design and implementation of performing and sustainable restoration and strengthening interventions that will ensure the integrity of the Katholikon and preserve its values.

References

1. Moropoulou, A., Mouzakis, Ch.: Chapter 1: Diagnostic study and assessment of the state of preservation of the Katholikon of Varnakova Monastery with nondestructive and analytical methods – proposal for compatible restoration materials. In: Interdisciplinary Diagnostic Study and Proposals for the Rehabilitation of the Katholiko and the Monastic Cells of the Varnakova Monastery in Fokida, Technical Report NTUA (2020). [in Greek]
2. Alexakis, E., Delegou, E.T., Lampropoulos, K.C., Apostolopoulou, M., Ntoutsi, I., Moropoulou, A.: NDT as a monitoring tool of the works progress and the assessment of materials and rehabilitation interventions at the Holy Aedicule of the Holy Sepulchre. Constr. Build. Mater. **189**, 512–526 (2018)
3. Daniels, D.J.: Ground Penetrating Radar, 2nd ed., Radar, Sonar, Navigation and Avionics Series 15, Institute of Electrical Engineers, London, UK (2004)
4. Jol, H.J. (ed.): Ground Penetrating Radar. Theory and Applications, 1st ed. Elsevier (2008)
5. Mpletsas, G.: Documentation and evaluation with nondestructive methods of the structural elements of the Monastery of Panagia Varnakova at Efpalio Doridos and Finite Element Models. MSE thesis, Interdisciplinary Postgraduate Program "Protection of Monuments", National Technical University of Athens, Thesis Supervisor Prof. A. Moropoulou (2019)

6. Moropoulou, A., Mouzakis, Ch.: Chapter 2: Evaluation of the mechanical characteristics of the masonries of the Katholikon and the monastic cells with the method of flat-jacks. In: Interdisciplinary Diagnostic Study and Proposals for the Rehabilitation of the Katholiko and the Monastic Cells of the Varnakova Monastery in Fokida, Technical Report NTUA (2020). [in Greek]
7. Moropoulou, A., Mouzakis, Ch.: Chapter 3: Seismic risk assessment and selection of time-histories for the seismic analysis of the Monastery buildings. In: Interdisciplinary Diagnostic Study and Proposals for the Rehabilitation of the Katholiko and the Monastic Cells of the Varnakova Monastery in Fokida, Technical Report NTUA (2020). [in Greek]
8. Comité Européen de Normalisation, CEN: Eurocode 8 (EC8-1): Design of structures for earthquake resistance Part 1: General rules, seismic actions and rules for buildings. CEN, Brussels (2004)
9. Moropoulou, A., Mouzakis, Ch: Chapter 4: Determination of the dynamic characteristics of the buildings of the Monastery of Varnakova with the method of microvibrations. In: Interdisciplinary Diagnostic Study and Proposals for the Rehabilitation of the Katholiko and the Monastic Cells of the Varnakova Monastery in Fokida, Technical Report NTUA (2020). [in Greek]

High Temperatures Impact on the Durability of Natural Stones: An Assessment by Means of Ultrasound Pulse Velocity Measurements

Theodore Bris[1], Ekaterini Delegou[1], Matteo Morabito[2], Elisabeta Zendri[3](✉), and Antonia Moropoulou[1](✉)

[1] Laboratory of Materials Science and Engineering, School of Chemical Engineering, National Technical University of Athens (NTUA), 9 Iroon Polytechniou Street, Zografou Campus, 15780 Athens, Greece
{edelegou,amoropul}@central.ntua.gr
[2] Arcadia Ricerche s.r.l, Venice, Italy
morabito@arcadiaricerche.eu
[3] Department of Environmental Sciences, Informatics and Statistics, Ca' Foscari University of Venice, Venice, Italy
elizen@unive.it

Abstract. It is common knowledge that wildfires are one of the most severe threats to cultural heritage sites, especially in conditions of climate change. This work discusses the results from tests which were performed in order to study the effects of fire on the strength and cohesion of natural building stones. Particularly, the Italian natural building stones of Istria, Carrara and Lecce were studied, because of their extensive use in the construction and decoration of important cultural heritage sites of the Italian peninsula. One of the most common methods to simulate the fire event is by exposing the specimens on increased temperatures within ovens. In this study, specimens of the three above mentioned lithotypes are exposed to the different high temperatures of 300 °C, 400 °C, 600 °C and 900 °C in an oven for 6 h. The changes induced by thermal shock were assessed, by measuring ultrasound pulse velocity (US) before and after treatment. It was observed that the increase of the exposure temperature led to a reduction of ultrasound pulse velocity for all the specimens' lithotypes. However, ultrasound pulse velocity decreased to different levels of the initial value for each lithotype per temperature interval. Thus, Istria stone demonstrated the greater reduction of ultrasound pulse velocity in the temperature interval of 300 °C–400 °C, Carrara marble in the temperature range of 400 °C–600 °C and Lecce stone in the temperature interval of 600 °C–900 °C, where it collapsed and lost completely its cohesion. It is concluded that Lecce stone after exposing to high temperatures presented the highest deterioration among the investigated lithotypes.

Keywords: Fire · Susceptibility to damage · Carbonate rocks · US · Physico-mechanical properties · Fire-related decay · Istria stone · Carrara marble · Lecce stone · Cultural heritage · NDT · Non-destructive testing

A. Osman and A. Moropoulou (Eds.): *Advanced Nondestructive and Structural Techniques for Diagnosis, Redesign and Health Monitoring for the Preservation of Cultural Heritage*, SPM 16, pp. 96–114, 2022.
https://doi.org/10.1007/978-3-031-03795-5_15

1 Introduction

The field of conservation and restoration of cultural heritage sites has always been crucial, mainly due to its importance of maintaining our memory of the past [1]. Especially nowadays, when a lot of catastrophic events, such as wildfires and floods, threaten not only cultural assets but also human lives, the study of the effects of these events and the attempt of protecting and preserving cultural heritage is imperative [2]. For the last decades, the study of the effects of fire has been a special issue for scientists, because of its increased presence [2–4]. Some cultural heritage sites have been damaged by fires (i.e. Notre Dame [5], Varnakova Monastery cells [6], La Fenice Theatre [7]), a phenomenon that is expected to enhance due to climate change [8]. Therefore, the investigation of the fire impact on building materials is getting rather important nowadays.

Many scientific studies have been focused on the simulation of fire and the study of its effects on the physical, chemical, and mechanical properties of stones [9]. These studies include mostly in-lab simulations of fire, through the exposure of stones' quarry specimens into an oven at different high temperatures [2, 10]. Some simulations take into account ash and fumes production as well, such as wood fire application [11]. Finally, real-world fire studies have been performed to a lesser extent [12], as well as laser-based techniques [11] in order to study the effect of high temperatures and related chemical reactions on stones.

In this study, Istria stone, Carrara marble, and Lecce stone have been studied because of their extensive use in the construction and decoration of several cultural heritage sites. Ducal Palace (Istria stone) [13], façade of the Certosa of Pavia (Carrara marble) [14], and Otranto Cathedral and Santa Caterina church in Galatina (Lecce stone) [15], are some of the most remarkable monuments where the investigated stones have been used. Taking into consideration the fire phenomenon outspread, the study of its impact on the above-mentioned stones is necessary.

Some references have already studied the fire effects on some of these stones. Especially as regards Carrara marble, some literature has focused on the mechanical changes that occur after fire [16], while others have connected these results with the chemical and physical changes in the marble [2, 10, 17]. Regarding Lecce stone, most of the previous works have mainly aimed to assess its general deterioration problems [15] as well as its behavior against moisture [18], when a small number of research have investigated the chemical and mechanical changes that take place on Lecce stone after fire [19]. As far as we know, the impact of fire on Istria stone has not been studied yet. The most of research has focused on the degradation that occurs by environmental factors, such as moisture and air pollution [20, 21] and further investigation is required to study the effects of fire on these stones.

Within this frame of reference, this paper investigates the effect of fire on the strength and durability of three Italian natural building stones and more specifically Istria stone, Carrara marble, and Lecce stone. Hereby, thermal treatments were applied by exposing the specimens of these three lithotypes to different high temperatures of 300 °C, 400 °C, 600 °C, and 900 °C in an oven for 6 h. Mechanical and physical damage to the stones was studied through an in-lab application of ultrasound pulse velocity measurements on selected cubic samples of each lithotype. This non-invasive technique is widely used in stone conservation to evaluate the efficacy of several stone treatments [1]. US measurements are also found in the literature for evaluating the effects of fire simulation experiments on stone specimens [2, 10, 22] as well as on in situ evaluation of post-fire consequences on masonries [23].

2 Materials and Methods

2.1 Stone Materials

Istria stone

Istria stone is one of the most common building stones in Venice, thanks to its special features, which make it particularly durable against seawater effects. It is being transferred to the city of Venice from the western side of the Istria peninsula, since the second half of the thirteenth century. The most important quarries for its extraction are Vrsar and Rovinj in Croatia, which is also the most popular and well-preserved [20].

From a petrographic point of view, Istria stone (Pietra d' Istria) is characterized as a micritic limestone [24] or mudstone [25]. It mostly contains: CaO: 97.50%, MgO: 0.08%, SiO_2: 1.02%, Fe_2O_3: 0.43% and Al_2O_3: 0.98% [21].

From a physic-mechanical point of view, the most important property of Istria limestone is its very low porosity and consequently its high coherence. These properties in collaboration with the small percentages of fine pores, classify the Istria stone as a very compact and frost resistant rock, making it the ideal barrier against seawater rising damp phenomena [20].

Carrara Marble

The Apuan Alps extracted Carrara marble is a type of white marble, popular in the field of sculpture and building decoration [26]. Cultural heritage owes to this Italian marble the existence of many important sculptures such as David and the Pietà di Michelangelo, respectively in Florence and Rome, and other historical buildings as the Pisa Tower, the façade of the Church of Santa Maria di Nazareth in Venice [27, 28].

Carrara marble, like other types of marble, is a metamorphic rock, consisting of fine calcite crystals. It is a homogeneous marble with a white to gray color, shiny fine grains, and gray veins. Except for calcite, which excels, traces of dolomite, quartz, plagioclase, and epidote have been detected as well [29] (mainly Ca, Mg, and less Sr [30]). A constant grain size distribution is demonstrated, on the order of hundreds of microns [29]. The development of small pores appears through the grain boundaries [16]. Thanks to its low porosity values and its high purity, it is largely used not only in construction and decoration but also in several experiments which study rock deformation [17].

Lecce Stone

In southern Italy, Lecce stone is widely used for both large-scale construction and decoration purposes. From Archaic to Roman ages, Lecce stone has decorated a lot of important monuments. It was used also later to build significant Romanesque churches during the middle Ages, as well as fortification towers in the Renaissance period. Nowadays, Lecce stone is extracted from the Cursi-Melpignano region (south of Lecce) [15].

Lecce stone belongs to the calcarenite family (calcareous sandstone [18] or wackestone [25]) which dates from the Miocene period. It is a fine-grained and homogeneous stone that mainly consists of calcium carbonate in the form of granules (microfossils, fossil fragments of marine life, and pellets) and calcite cement [18]. A typical composition of Lecce stone contains mainly calcite with low magnesium fragments. Other often observed components are clay minerals (glauconite, illite, kaolinite, smectite, and chlorite) and a fraction of silicate minerals such as quartz, micas, and pyroxenes [15].

Lecce stone has a characteristic pale yellow color and demonstrates high porosity and consequently low strength which make it a sensitive material to decay [15]. However, its availability and its workability are determinant factors that justify its extensive use [1].

2.2 Experimental Procedure

The fire was simulated by exposing the three lithotypes in four different high temperatures (300 °C, 400 °C, 600 °C, and 900 °C) for six hours in an oven. Twenty-four cubic specimens (5 cm × 5 cm × 5 cm) of Istria and twenty-four cubic specimens of Lecce stones were studied in total before and after treatment, whereas three reference specimens of each stone were studied as well. Eighteen cubic specimens of Carrara marble were studied before and after treatment, as well as three reference specimens were investigated. Six specimens of each stone were examined at each temperature, except from Carrara marble of 300 °C and 400 °C temperatures, where three specimens were studied.

After drying to a stable weight at room temperature, ultrasound pulse velocity was measured for each specimen. After treatment, the specimens were submitted to the same procedure in order to record the changes.

2.3 Techniques

To investigate the change of physical and mechanical behavior of the three lithotypes, ultrasound pulse (US) velocity was measured by using the direct transmission method. Travel time of the pulse through the stone was determined by using a Pundit tester along the core's axis of the specimens and probes (two transducers-a transmitters and a receiver) with a frequency of 54 kHz. Lubricant material, covered with a special film, was used as a binder in order to enhance the transducer-sample contact. Velocities were calculated by dividing the length of the sample by the total travel time of the pulse, which has been measured. In total, nine consecutive, one-direction measurements were performed on each specimen. In particular, three measurements were operated at three points of each specimen; 0,25 cm (US1), 1,25 cm (US2), and 2,5 cm (US3) from the edge, respectively, as shown in Fig. 1.

Fig. 1. Three points on specimen where the measurements of US velocity were performed.

The mean value of the US velocity for each point (height) was calculated. Subsequently, the US velocity value of each specimen was calculated by the respective values of the three points, as well as ultrasound pulse velocities of the total of the specimens per lithotype were averaged. All velocity measurements were performed in like manner before and after thermal treatment.

3 Results and Discussion

3.1 Investigation of the Effects of Fire Simulation on Istria Stone

In Fig. 2, the Istria stone reference specimen and the specimens after the exposure at high temperatures are presented. After the exposure at 300 °C and 400 °C, no significant color alterations are observed, while after the exposure at 600 °C and 900 °C, the most important deterioration is achieved. Particularly, at 600 °C color changes are presented and at 900 °C cracks and loss of the original material are observed.

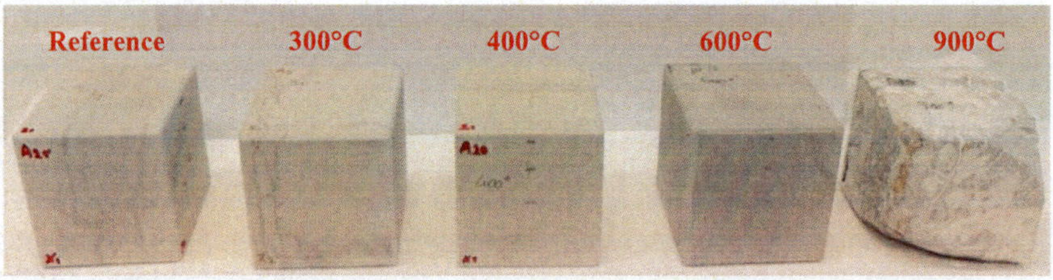

Fig. 2. Istria stone reference specimen and specimens after fire simulation.

In Table 1, ultrasound pulse velocity for each point, the mean value of US velocity and standard deviation per Istria stone specimen (I) are reported.

It is observed that the velocities measured before heat treatment are repeatable, which proves the homogeneity of Istria stone [20]. As can be seen, the US velocity values for each point do not change significantly as we move to the center of the specimen either before or after thermal treatment. This fact is also confirmed by the standard deviation, which remains at most cases below ±50 m/s. There are some cases though at 600 °C (I9) and at 900 °C (I12, I13, and I16), where the standard deviation is above ±100 m/s. As regards the I16, the US velocity is lower at the edge than the center of the specimen, which demonstrates that the impact of fire simulation is more significant to the area

Table 1. US velocity values per specimen of Istria stone

Stone sample code	Temperature (°C)	US1 velocity (m/s)	US2 velocity (m/s)	US3 velocity (m/s)	US velocity mean value (m/s)	Std. deviation (m/s)
I1r	Reference	2669,0	2673,8	2717,4	2686,7	±26,6
I2r		2762,4	2727,3	2697,8	2729,2	±32,3
I3r		2681,8	2715,7	2701,1	2699,5	±17,0
I1	BT*	2725,8	2648,3	2643,6	2672,6	±46,1
	300 °C	2672,0	2701,1	2696,2	2689,8	±15,6
I2	BT	2708,1	2627,6	2547,5	2627,7	±80,3
	300 °C	2717,9	2752,7	2632,2	2701,0	±62,0
I3	BT	2649,6	2659,0	2682,8	2663,8	±17,1
	300 °C	2777,4	2726,8	2721,8	2742,0	±30,7
I4	BT	2659,0	2626,3	2673,2	2652,8	±24,0
	300 °C	2756,9	2726,8	2751,8	2745,2	±16,1
I5	BT	2571,4	2621,1	2621,1	2604,5	±28,7
	300 °C	2667,9	2653,6	2667,9	2663,1	±8,2
I6	BT	2648,9	2621,1	2653,6	2641,2	±17,6
	300 °C	2741,3	2716,4	2648,9	2702,2	±47,8
I19	BT	2655,3	2655,3	2650,5	2653,7	±2,7
	400 °C	2230,8	2071,4	2240,9	2181,0	±95,1
I20	BT	2733,7	2699,5	2723,8	2719,0	±17,6
	400 °C	2206,1	2113,4	2307,3	2209,0	±97,0
I21	BT	2730,3	2695,7	2657,1	2694,4	±36,6
	400 °C	2285,7	2204,4	2285,7	2258,6	±46,9
I22	BT	2707,6	2707,6	2707,6	2707,6	0,0
	400 °C	2293,6	2202,6	2293,6	2263,3	±52,5
I23	BT	2776,1	2676,3	2657,1	2703,2	±63,9
	400 °C	2275,2	2204,4	2285,7	2255,1	±44,2
I24	BT	2314,4	2326,4	2405,4	2348,7	±49,4
	400 °C	1943,7	1894,5	1935,3	1924,5	±26,3
I7	BT	2712,5	2688,2	2659,6	2686,7	±26,5
	600 °C	2202,6	2193,0	2212,4	2202,7	±9,7
I8	BT	2729,3	2704,4	2600,0	2677,9	±68,6
	600 °C	2041,3	2041,3	2016,3	2033,0	±14,4
I9	BT	2724,8	2670,9	2670,9	2688,8	±31,1
	600 °C	2133,6	1972,1	1644,5	1916,8	±249,2
I10	BT	2666,1	2642,3	2637,7	2648,7	±15,2
	600 °C	2079,8	2088,6	2088,6	2085,7	±5,1
I11	BT	2683,8	2683,8	2659,8	2675,8	±13,8
	600 °C	2206,7	2216,5	2206,7	2210,0	±5,7

(continued)

Table 1. (*continued*)

Stone sample code	Temperature (°C)	US1 velocity (m/s)	US2 velocity (m/s)	US3 velocity (m/s)	US velocity mean value (m/s)	Std. deviation (m/s)
I18	BT	2707,6	2688,2	2669,0	2688,3	±19,3
	600 °C	2055,1	1921,8	2006,0	1994,3	±67,4
I12	BT	2673,5	2692,5	2702,1	2689,4	±14,6
	900 °C	2167,4	1947,3	1902,7	2005,8	±141,7
I13	BT	2665,8	2656,3	2661,0	2661,0	±4,7
	900 °C	1887,8	1754,4	1644,0	1762,1	±122,1
I14	BT	2641,0	2604,0	2693,5	2646,2	±45,0
	900 °C	1945,5	1872,7	1976,3	1931,5	±53,2
I15	BT	2732,2	2673,8	2645,5	2683,8	±44,2
	900 °C	1864,2	2016,3	2041,3	1973,9	±95,9
I16	BT	2665,5	2660,7	2627,7	2651,3	±20,6
	900 °C	1655,6	2066,1	2314,8	2012,2	±332,9
I17	BT	2762,4	2727,3	2717,4	2735,7	±23,7
	900 °C	1953,1	1818,2	1938,0	1903,1	± 73,9

*BT: Before treatment.

close to the edge than the center of the stone. In reverse, concerning I9, I12 and I13, it is observed that US velocity decreases as we move to the center of the stone. This can be attributed to the presence of localized defaults such as veins and micro-cracks at the central area of these particular specimens, which behave differently at the high temperature impact.

The results of the ultrasound pulse velocity change per temperature for Istria stone are depicted graphically in Fig. 3. The mean values of US velocity for all the investigated Istria stone specimens before and after fire simulation, the percentage change of US velocity after thermal treatment, and the percentage change of the US velocity per temperature interval of Istria stone are presented in Table 2.

Table 2. Percentage change of US velocity per exposure temperature and per temperature interval for Istria stone

Stone sample code	US velocity (mean values) (m/s)		Percentage change of US velocity after thermal treatment	Percentage change of US velocity per temperature interval	
	Before (room temperature)	After (thermal treatment)		Temperature interval	%
$Istria_{300\ °C}$	2643,8 ± 25,0	2707,2 ± 31,5	+2,40%	25 °C–300 °C	+2,40%
$Istria_{400\ °C}$	2637,8 ± 143,4	2181,9 ± 130,3	−17,28%	300 °C–400 °C	−19,68%
$Istria_{600\ °C}$	2677,7 ± 15,2	2073,8 ± 116,6	−22,55%	400 °C–600 °C	−5,27%
$Istria_{900\ °C}$	2677,9 ± 33,2	1931,4 ± 93,1	−27,88%	600 °C–900 °C	−5,32%

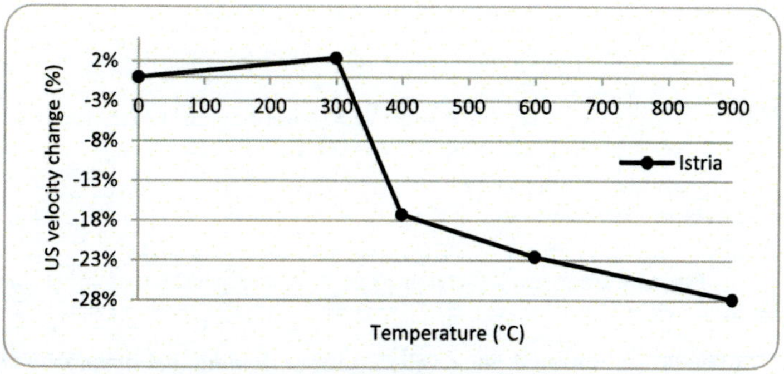

Fig. 3. Percentage change of US velocity related to temperature (°C) for Istria stone

As it is inferred, after the exposure of the Istria specimens at 300 °C, the US velocity values display a slight increase of 2.40%. This is probably due to the dilation of the calcite crystals against shrinkage of the low-content clay minerals [22]. Particularly, despite the compaction of the clay minerals found in colored veins of the stone, the simultaneous expansion of the fine-grained calcite crystals fills the voids and the micro-cavities of the low porosity stone [21, 22]. Consequently, after the exposure at 300 °C, the calcite expansion leads to the simultaneous enhancement of Istria stone.

In reverse, after the exposure of Istria stone at higher temperatures (400 °C, 600 °C, and 900 °C), the US velocities decreased with increasing temperature. More specifically, at 400 °C the lowest reduction of 17.28% is observed, while at 600 °C the percentage change of US velocity is −22.55%. Most of the damage on Istria stone occurred after its exposure at 900 °C since it demonstrates the highest reduction of the US velocity, which is equal to 27.88%. This is due to the cracking within grains and the separation of grain boundaries, which is more intense as the temperature increases [22]. These results indicate that the cohesion and the mechanical strength of Istria stone decreases with the oven temperature increase.

Regarding the percentage change of US velocity per temperature interval, it is observed that most damage in the consistency of Istria stone occurs in the temperature interval of 300 °C–400 °C, with a reduction of 19,68%. This indicates that at 400 °C, the separation along crystal boundaries and cracking within grains starts [22]. Subsequently, in 400 °C–600 °C the percentage change is −5.27% and in 600 °C–900 °C the reduction is 5.32%. The low changes of the velocity values in the last-mentioned temperature intervals confirm that most of the decrease on Istria stone's mechanical strength takes place in 300 °C–400 °C.

3.2 Investigation of the Effects of fire Simulation on Carrara Marble

In Fig. 4, the Carrara marble reference specimen and the specimens after the exposure at high temperatures are presented. It is observed that after the exposure at 300 °C there are no significant changes. In reverse, after the exposure at 400 °C, 600 °C, and 900 °C, Carrara marble specimens lose their geometry, as well as loss of original material is observed.

Fig. 4. Carrara marble reference specimen and specimens after fire simulation

In Table 3, ultrasound pulse velocity for each point, the mean value of US velocity and standard deviation per Carrara marble specimen (C) are reported.

The US velocity values for each point remain almost stable as we move to the center of the specimen, as well as the standard deviation does not go beyond ±50 m/s. Furthermore, it is observed that the velocities measured before heat treatment are repeatable, which proves the homogeneity of Carrara marble [29].

Table 3. US velocity values per specimen of Carrara marble

Stone sample code	Temperature (°C)	US1 velocity (m/s)	US2 velocity (m/s)	US3 velocity (m/s)	US velocity mean value (m/s)	Std. deviation (m/s)
C1r	Reference	2541,4	2495,0	2558,7	2531,7	±32,9
C2r		2566,3	2570,7	2531,9	2556,3	±21,3
C3r		2477,0	2464,8	2501,7	2481,1	±18,8
C7	BT*	2525,3	2495,8	2451,0	2490,7	±37,4
	300 °C	2118,6	2092,1	2124,6	2111,8	±17,3
C8	BT	2500,0	2447,0	2447,0	2464,7	±30,6
	300 °C	2193,0	2161,4	2155,2	2169,8	±20,3
C9	BT	2592,8	2557,6	2544,7	2565,0	±24,9
	300 °C	2090,0	2053,1	2033,7	2058,9	±28,6
C10	BT	2509,1	2476,3	2460,3	2481,9	±24,9
	400 °C	1807,1	1767,2	1769,2	1781,2	±22,5
C11	BT	2427,5	2479,4	2483,5	2463,5	±31,2
	400 °C	1767,3	1722,9	1711,1	1733,8	±29,6
C12	BT	2540,4	2506,6	2498,3	2515,1	±22,3
	400 °C	1798,6	1792,2	1775,3	1788,7	±12,0
C1	BT	2529,5	2487,6	2475,2	2497,4	±28,4
	600 °C	1204,8	1141,6	1171,0	1172,4	±31,7
C2	BT	2484,4	2472,2	2452,1	2469,6	±16,3
	600 °C	1066,7	1038,1	1082,8	1062,6	±22,6
C3	BT	2587,6	2518,4	2485,1	2530,4	±52,3
	600 °C	1331,6	1277,4	1310,7	1306,5	±27,3

(continued)

Table 3. (*continued*)

Stone sample code	Temperature (°C)	US1 velocity (m/s)	US2 velocity (m/s)	US3 velocity (m/s)	US velocity mean value (m/s)	Std. deviation (m/s)
C13	BT	2612,1	2559,1	2580,9	2584,0	±26,6
	600 °C	1166,3	1127,2	1169,0	1154,2	±23,4
C14	BT	2536,1	2465,7	2422,2	2474,7	±57,5
	600 °C	1248,1	1191,9	1212,0	1217,4	±28,5
C15	BT	2499,2	2466,4	2499,2	2488,2	±18,9
	600 °C	1097,4	1078,5	1145,0	1106,9	±34,3
C4	BT	2586,5	2517,5	2525,9	2543,3	±37,6
	900 °C	1025,5	999,0	964,6	996,3	±30,5
C5	BT	2582,1	2521,7	2534,4	2546,1	±31,8
	900 °C	1017,2	1038,1	1049,0	1034,8	±16,2
C6	BT	2507,5	2470,6	2466,6	2481,5	±22,5
	900 °C	1024,4	998,0	936,8	986,4	±44,9
C16	BT	2564,8	2509,2	2517,6	2530,5	±30,0
	900 °C	1039,4	1128,4	1077,4	1081,7	±44,6
C17	BT	2518,3	2473,0	2453,0	2481,4	±33,5
	900 °C	1113,7	1192,7	1089,6	1132,0	±53,9
C18	BT	2530,3	2496,7	2443,9	2490,3	±43,6
	900 °C	1072,8	1096,3	1035,1	1068,1	±30,9

*BT: Before treatment.

The results of the ultrasound pulse velocity change per temperature for Carrara marble are depicted graphically in Fig. 5, while the mean values of US velocity for all the investigated Carrara marble specimens before and after fire simulation, the percentage change of US velocity after thermal treatment, and the percentage change of the US velocity per temperature interval of Carrara marble are presented in Table 4.

Table 4. Percentage change of US velocity per exposure temperature and per temperature interval for Carrara marble

Stone sample code	US velocity (mean values) (m/s)		Percentage change of US velocity after thermal treatment	Percentage change of US velocity per temperature interval	
	Before (room temperature)	After (thermal treatment)		Temperature interval	%
$Carrara_{300°C}$	2506,8 ± 52,1	2113,5 ± 55,5	−15,69%	25 °C–300 °C	−15,69%
$Carrara_{400°C}$	2486,8 ± 26,2	1767,9 ± 29,8	−28,91%	300 °C–400 °C	−13,22%
$Carrara_{600°C}$	2507,4 ± 43,3	1170,0 ± 85,6	−53,34%	400 °C–600 °C	−24,43%
$Carrara_{900°C}$	2512,2 ± 31,1	1049,9 ± 55,2	−58,21%	600 °C–900 °C	−4,87%

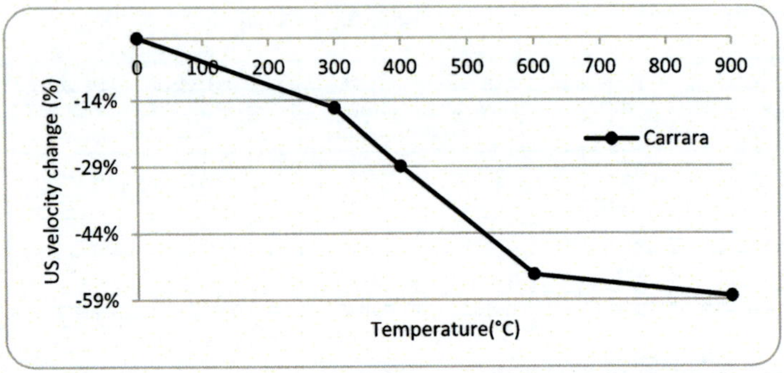

Fig. 5. Percentage change of US velocity related to temperature (°C) for Carrara marble

As can be seen, after the exposure of Carrara marble to high temperatures (300 °C, 400 °C, 600 °C, and 900 °C) the US velocity values decrease. Moreover, it is observed that the higher the exposure temperature of the specimens, the lower the US velocity. In particular, at 300 °C the velocity values display the lowest reduction of 15.69%, while at 400 °C the decrease is equal to 28.91%. The next higher decline of 53.34% is recorded after the exposure at 600 °C, while the highest reduction of the US velocities is noticed at 900 °C and is equal to 58.21%. The extremely high decrease of the US velocities at 600 °C and 900 °C, is probably due to the differential expansion of calcite crystals that occurs in temperatures above 600 °C and provokes complete disruption [31]. As a result, it is concluded that the cohesion and the mechanical strength of Carrara marble decrease with the exposing temperature increase.

Concerning the percentage change of US velocity per temperature interval, the most significant damage in the durability of Carrara marble occurs in the temperature interval of 400 °C–600 °C, with a reduction of 24.43%. This proves that following this threshold temperature of 600 °C important damage started in Carrara marble, because of the extensive distortion of calcite crystals [22]. Then, the second higher decrease of the US velocity values appears in the temperature interval of 25 °C–300 °C with a reduction of 15.69%. US velocity reduction in this temperature interval can be attributed to a non-reversible calcite expansion, known as thermal hysteresis [31]. In the temperature interval of 300 °C–400 °C, a low decline of 13.22% is observed and finally, the lowest reduction is recorded in 600 °C–900 °C. The last-mentioned insignificant decrease proves that most of the damage has been already induced from the previous temperature intervals and especially in 400 °C–600 °C.

3.3 Investigation of the Effects of Fire Simulation on Lecce Stone

In Fig. 6, the Lecce stone reference specimen and the specimens after fire simulation are presented. After the exposure at 300 °C and 400 °C, no notable alterations are observed. However, after the exposure at 600 °C, color changes are presented, while at 900 °C Lecce stone fully collapses.

Fig. 6. Lecce stone reference specimen and specimens after fire simulation

In Table 5, ultrasound pulse velocity for each point, the mean value of US velocity and standard deviation per Lecce stone specimen (L) are reported.

As can be seen, the US velocity values for each point do not change significantly as we move to the center of the specimen either before or after thermal treatment, which is confirmed by the low values of the standard deviation (below ±50 m/s) as well. However, it is observed that the velocities, measured before fire simulation, show lower repeatability compared to the other two lithotypes, probably due to Lecce stone porous microstructure [1], without excluding the possible compositional differences among the investigated specimens (differences in clay and silicate minerals fractions [15]). In some cases after the exposure at 600 °C (L13, L17 and L18) the standard deviation of the recorded US velocity values is above ±100 m/s. In particular, the US velocity decreases as we move to the center of the specimens, which is probably due to the differences of Lecce stone microstructure and clay/silicate minerals amount.

Finally, it becomes clear that the specimens exposed at 900 °C were not investigated by US means, because they collapsed after thermal treatment.

Table 5. US velocity values per specimen of Lecce stone

Stone sample code	Temperature (°C)	US1 velocity (m/s)	US2 velocity (m/s)	US3 velocity (m/s)	US velocity mean value (m/s)	Std. deviation (m/s)
L1r	Reference	2035,1	2021,5	1992,1	2016,2	±22,0
L2r		2145,7	2140,0	2148,6	2144,8	±4,4
L3r		1934,0	1948,8	1936,5	1939,8	±8,0
L1	BT*	2052,0	2005,3	2013,4	2023,6	±24,9
	300 °C	2018,8	1965,9	1978,9	1987,9	±27,6
L2	BT	1858,4	1875,0	1849,1	1860,8	±13,1
	300 °C	1786,1	1794,9	1797,1	1792,7	±5,8
L3	BT	2054,1	2032,1	2029,4	2038,6	±13,6
	300 °C	1984,3	1968,9	1984,3	1979,2	±8,9
L4	BT	2037,9	1992,1	1968,6	1999,5	±35,2
	300 °C	1963,5	1948,3	1963,5	1958,4	±8,8
L5	BT	1872,2	1849,0	1835,4	1852,2	±18,6
	300 °C	1830,9	1802,2	1791,4	1808,1	±20,4

(continued)

Table 5. (*continued*)

Stone sample code	Temperature (°C)	US1 velocity (m/s)	US2 velocity (m/s)	US3 velocity (m/s)	US velocity mean value (m/s)	Std. deviation (m/s)
L6	BT	2039,6	2020,6	2004,6	2021,6	±17,5
	300 °C	2047,8	2036,9	2036,9	2040,5	±6,3
L7	BT	2010,6	1997,4	2015,9	2008,0	±9,6
	400 °C	1855,7	1853,5	1849,0	1852,7	±3,5
L8	BT	2146,5	2128,5	2107,9	2127,6	±19,3
	400 °C	1905,0	1876,8	1845,0	1875,6	±30,0
L9	BT	2043,2	2021,4	1989,5	2018,0	±27,0
	400 °C	1852,9	1848,4	1873,6	1858,3	±13,4
L10	BT	2062,2	2037,2	2028,9	2042,8	±17,3
	400 °C	1884,4	1845,2	1856,5	1862,0	±20,2
L11	BT	2046,9	2030,6	2058,0	2045,2	±13,8
	400 °C	1848,7	1800,7	1794,4	1814,6	±29,7
L12	BT	2069,4	2033,4	2033,4	2045,4	±20,8
	400 °C	1823,7	1791,5	1815,0	1810,1	±16,7
L13	BT	1779,6	1752,9	1798,6	1777,0	±23,0
	600 °C	1249,4	1144,8	925,0	1106,4	±165,5
L14	BT	2042,5	2023,4	2020,7	2028,9	±11,9
	600 °C	1492,6	1413,2	1458,1	1454,6	±39,8
L15	BT	2090,0	2064,3	2055,9	2070,1	±17,8
	600 °C	1538,2	1488,2	1501,5	1509,3	±25,9
L16	BT	2059,2	2031,5	2053,6	2048,1	±14,6
	600 °C	1538,1	1497,0	1542,8	1526,0	±25,2
L17	BT	1871,7	1871,7	1832,9	1858,8	±22,4
	600 °C	1216,0	1154,4	891,5	1087,3	±172,4
L18	BT	2068,4	2037,7	2043,2	2049,8	±16,3
	600 °C	1354,8	1100,4	938,5	1131,3	±209,9
L19	BT	2000,0	1994,7	1992,1	1995,6	±4,0
	900 °C	x**	x	x	x	x
L20	BT	1933,2	1867,2	1923,5	1908,0	±35,6
	900 °C	x	x	x	x	x
L21	BT	2069,5	2061,0	2063,8	2064,7	±4,3
	900 °C	x	x	x	x	x
L22	BT	2128,7	2084,7	2099,2	2104,2	±22,4
	900 °C	x	x	x	x	x
L23	BT	2120,1	2062,5	2059,7	2080,8	±34,1
	900 °C	x	x	x	x	x
L24	BT	2012,0	2006,6	1977,8	1998,8	±18,4
	900 °C	x	x	x	x	x

*BT: Before treatment, **x: collapsed specimen.

The mean values of US velocity for all the investigated Lecce stone specimens before and after fire simulation, the percentage change of US velocity after thermal treatment, and the percentage change of the US velocity per temperature interval of Lecce stone are recorded in Table 6, while the results of the ultrasound pulse velocity change per temperature for Lecce stone are presented graphically in Fig. 7.

Table 6. Percentage change of US velocity per exposure temperature and per temperature interval for Lecce stone

Stone sample code	US velocity (mean values) (m/s)		Percentage change of US velocity after thermal treatment	Percentage change of US velocity per temperature interval	
	Before (room temperature)	After (thermal treatment)		Temperature interval	%
$Lecce_{300°C}$	1966,1 ± 85,8	1927,8 ± 102,4	−1,95%	25 °C–300 °C	−1,95%
$Lecce_{400°C}$	2047,8 ± 42,1	1845,6 ± 26,8	−9,88%	300 °C–400 °C	−7,93%
$Lecce_{600°C}$	1972,1 ± 122,9	1302,5 ± 214,4	−33,96%	400 °C–600 °C	−24,08%
$Lecce_{900°C}$	2025,4 ± 72,4	0	−100,00% (collapse)	600 °C–900 °C	−66,04%

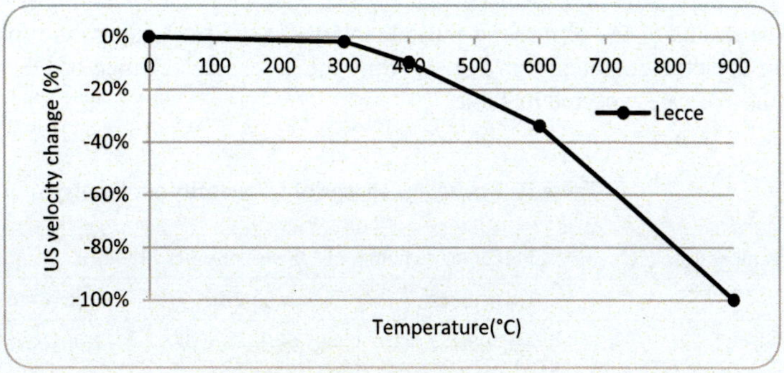

Fig. 7. Percentage change of US velocity related to temperature (°C) for Lecce stone

The US velocity values, and so the durability, of Lecce stone decrease with the increase of the exposure temperature. In particular, at 300 °C the velocity values demonstrate the lowest reduction of 1.95%, while at 400 °C the reduction is equal to 9.88%. The low percentage change values after the last-mentioned temperatures declare that the stone susceptibility to fire impact is not significant. This can be possibly attributed to the shrinkage of clay minerals against the expansion of calcite [22]. The next higher decline of US velocity values is observed after the exposure at 600 °C (−33.96%). The high cohesion decrease is probably due to the presence of quartz in Lecce stone. More specifically, quartz demonstrates internal disintegration at temperatures above 573 °C, resulting in weakening [31]. The highest reduction of the US velocities is noticed after the exposure at 900 °C, where Lecce specimens collapse, and the percentage change is considered −100.00%. This is probably due to the presence of some clay minerals in the lithotype [15], such as kaolinite and chlorite, which decompose at temperatures above 600 °C [31]. Equally significant is the differential thermal expansion of calcite that occurs at high temperatures (>600 °C). This fact increases the tensile and compressive stresses, resulting in permanent strain in the stone.

In regard to the percentage change of US velocity per temperature interval, most of the damage in Lecce stone occurs in the temperature interval of 600 °C–900 °C, with a reduction of 66.04%. At the threshold temperature of 600 °C a significant damage in Lecce stone is commenced, due to the complete expansion of clay and silicate minerals [31]. The second higher decrease of the US velocity values appears in the temperature interval of 400 °C–600 °C by 24.08%. As has been already mentioned, quartz grains suffer from internal rupturing in this temperature interval, which can induce strength decrease and friability [31]. A low decline is observed (7.93%) in the temperature interval of 300 °C–400 °C and finally, the lowest reduction is recorded in 25 °C–300 °C interval (1.95%). The low reduction values in the two last-mentioned temperature intervals indicate that no significant damage in the consistency and the mechanical strength of Lecce stone occurs at temperatures below 400 °C; whereas significant damages are evident in temperatures above 600 °C.

3.4 Comparative Study

The results of the ultrasound pulse velocity change per temperature for all lithotypes are depicted graphically in Fig. 8, while the percentage change of US velocity for all lithotypes are reported in Table 7.

Table 7. Percentage change of US velocity per lithotype

Temperature (°C)	Percentage change of US velocity per lithotype		
	Istria stone	Carrara marble	Lecce stone
300 °C	+2,40%	−15,69%	−1,95%
400 °C	−17,28%	−28,91%	−9,88%
600 °C	−22,55%	−53,34%	−33,96%
900 °C	−27,88%	−58,21%	−100,00% (collapse)

According to Fig. 8 and Table 7, it is observed that Istria stone is more durable than the other two investigated lithotypes, as it demonstrates the lowest reduction of the US velocity after the fire simulation. Then, Carrara marble displays higher durability (−58.21%) compared to Lecce stone, which eventually shows the higher susceptibility to the fire impact, with a reduction of the US velocity by 100.00%. It is worth mentioning that Carrara marble shows the higher decline of the US velocity at temperatures below

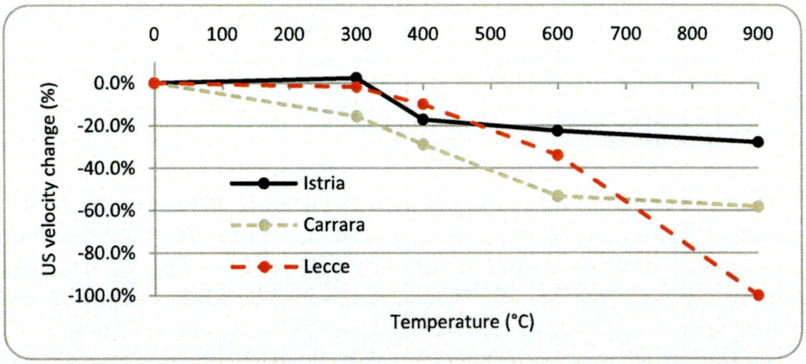

Fig. 8. US velocity percentage change related to temperature (°C) for all lithotypes

600 °C; higher even from Lecce stone, which finally collapses at the temperature of 900 °C.

The percentage change of US velocity per temperature interval for each lithotype is presented in Table 8.

Table 8. Percentage change of US velocity per temperature interval for each lithotype

Temperature interval	Percentage change of US velocity per lithotype per temperature interval		
	Istria stone	Carrara marble	Lecce stone
25 °C–300 °C	+ 2,40%	−15,69%	−1,95%
300 °C–400 °C	−19,68%	−13,22%	−7,93%
400 °C–600 °C	−5,27%	−24,43%	−24,08%
600 °C–900 °C	−5,32%	−4,87%	−66,04%

As can be seen from Table 8, in the temperature interval of 25 °C–300 °C, Carrara marble shows the highest reduction of the US velocity (15.69%); temperature range where thermal hysteresis occurs [31]. Istria stone demonstrates the highest decrease of the velocity values compared to the other lithotypes (19.68%), in the temperature interval of 300 °C–400 °C, where separation along crystal boundaries and cracking within grains start [22]. In the temperature interval of 400 °C–600 °C, most of the damage occurs in Carrara marble and Lecce stone, which demonstrate a reduction of 24.43% and 24.08% respectively. In this temperature interval, concerning Carrara marble, the complete expansion of calcite crystals takes place [22], while in Lecce stone the quartz fractions start to disintegrate [31]. Finally, in the temperature interval of 600 °C–900 °C, Lecce stone shows the highest reduction of the US velocity with a value of 66.04%.

It is also worth noting that the highest damage is induced at different temperature intervals for each lithotype under investigation. In particular, the most of damage in Istria stone takes place in the temperature interval of 300 °C–400 °C, while in Carrara marble occurs in 400 °C–600 °C. Finally, Lecce stone displays the highest loss of its cohesion in the temperature interval of 600 °C–900 °C, where it collapses.

4 Conclusions

During this preliminary study, the effects of high temperatures on the durability of three natural stones (Istria stone, Carrara marble, and Lecce stone), used extensively in the construction and decoration of many cultural heritage sites, are investigated. In this work, the fire event was simulated by exposing the three lithotypes to different high temperatures (300 °C, 400 °C, 600 °C, and 900 °C) for 6 h, in an oven.

Ultrasound pulse velocity (US) measurements proved to be an essential non-destructive technique for the physical and mechanical characterization of the stones. It is observed that the increase of the exposure temperature on the samples of all lithotypes leads to the reduction of the US velocity values, a fact that implies the decrease of the cohesion and mechanical strength of the stones. However, each lithotype demonstrates different percentage changes of the US velocity values per temperature interval. In particular, Istria stone shows the highest reduction of its consistency in the temperature interval of 300 °C–400 °C, while Carrara marble's most significant damage occurs in 400 °C–600 °C. Finally, Lecce stone demonstrates the highest decrease in the US velocities, so as in its mechanical strength, in the temperature interval of 600 °C–900 °C.

To conclude, the highest deterioration after the exposure at high temperatures is shown by Lecce stone, which is the only lithotype that collapsed after the exposure at 900 °C. Then, Carrara marble demonstrates the second lower durability to the impact of high temperatures, while Istria stone is the most durable one compared to the other two investigated lithotypes.

References

1. Vasanelli, E., Colangiuli, D., Calia, A., Sileo, M., Aiello, M.A.: Ultrasonic pulse velocity for the evaluation of physical and mechanical properties of a highly porous building limestone. Ultrasonics **60**, 33–40 (2015). https://doi.org/10.1016/j.ultras.2015.02.010
2. Brotóns, V., Tomás, R., Ivorra, S., Alarcón, J.C.: Temperature influence on the physical and mechanical properties of a porous rock: San Julian's calcarenite. Eng. Geol. **167**, 117–127 (2013). https://doi.org/10.1016/j.enggeo.2013.10.012
3. Vigroux, M., Eslami, J., Beaucour, A.L., Bourgès, A., Noumowé, A.: High temperature behaviour of various natural building stones. Constr. Build. Mater. **272** (2021). https://doi.org/10.1016/j.conbuildmat.2020.121629
4. Sciarretta, F., Eslami, J., Beaucour, A.-L., Noumowé, A.: State-of-the-art of construction stones for masonry exposed to high temperatures. Constr. Build. Mater. **304**, 124536 (2021). https://doi.org/10.1016/J.CONBUILDMAT.2021.124536
5. Deldicque, D., Rouzaud, J.N.: Temperatures reached by the roof structure of Notre-Dame de Paris in the fire of April 15th 2019 determined by Raman paleothermometry. Comptes Rendus Geosci. **352**(1), 7–18 (2020). https://doi.org/10.5802/CRGEOS.9
6. Delegou, E.T., et al.: The effect of fire on building materials: the case-study of the Varnakova Monastery cells in central Greece. Heritage **2**(2), 1233–1259 (2019). https://doi.org/10.3390/heritage2020080

7. Russo, S., Sciarretta, F.: Masonry exposed to high temperatures: mechanical behaviour and properties - an overview. Fire Saf. J. **55** (2013). https://doi.org/10.1016/j.firesaf.2012.10.001
8. Harrison, S.P., Marlon, J.R., Bartlein, P.J.: Changing climates, earth systems and society. Chang. Clim. Earth Syst. Soc. (2010). https://doi.org/10.1007/978-90-481-8716-4
9. Martinho, E., Dionísio, A.: Assessment techniques for studying the effects of fire on stone materials: a literature review. Int. J. Archit. Herit. **14**(2), 275–299 (2020). https://doi.org/10.1080/15583058.2018.1535008
10. Murru, A., Freire-Lista, D.M., Fort, R., Varas-Muriel, M.J., Meloni, P.: Evaluation of post-thermal shock effects in Carrara marble and Santa Caterina di Pittinuri limestone. Constr. Build. Mater. **186**, 1200–1211 (2018). https://doi.org/10.1016/j.conbuildmat.2018.08.034
11. McCabe, S., Smith, B.J., Warke, P.A.: Exploitation of inherited weakness in fire-damaged building sandstone: the 'fatiguing' of 'shocked' stone. Eng. Geol. **115**(3–4), 217–225 (2010). https://doi.org/10.1016/j.enggeo.2009.06.003
12. Chakrabarti, B., Yates, T., Lewry, A.: Effect of fire damage on natural stonework in buildings. Constr. Build. Mater. **10**(7) (1996). https://doi.org/10.1016/0950-0618(95)00076-3
13. Thorn, A.: 12th International Congress on the Deterioration and Conservation of Stone Columbia University, New York, 2012. In: Lithium Silicate Consolidation of Wet Stone and Plaster, pp. 0–10 (2012)
14. Bruni, S., Cariati, F., Bianchi, C.L., Zanardini, E., Sorlini, C.: Spectroscopic investigation of red strains affecting the Carrara marble facade of the certosa of Pavia. Archaeometry **37**(2) (1995). https://doi.org/10.1111/j.1475-4754.1995.tb00741.x
15. Calia, A., Tabasso, M.L., Mecchi, A.M., Quarta, G.: The study of stone for conservation purposes: Lecce stone (southern Italy). Geol. Soc. Spec. Publ. **391**(1), 139–156 (2014). https://doi.org/10.1144/SP391.8
16. Malaga-Starzec, K., Lindqvist, J.E., Schouenborg, B.: Experimental study on the variation in porosity of marble as a function of temperature. Geol. Soc. Spec. Publ. **205**, 81–88 (2002). https://doi.org/10.1144/GSL.SP.2002.205.01.07
17. Rutter, E.H.: Experimental study of the influence of stress, temperature, and strain on the dynamic recrystallization of Carrara marble. J. Geophys. Res. **100**(B12), 651–663 (1995). https://doi.org/10.1029/95jb02500
18. D'Agostino, D.: Moisture dynamics in an historical masonry structure: The Cathedral of Lecce (South Italy). Build. Environ. **63**, 122–133 (2013). https://doi.org/10.1016/j.buildenv.2013.02.008
19. Vasanelli, E., Quarta, G., Micelli, F., Calia, A.: The effects of an historical fire on a porous calcarenite from an industrial-archaeological building in the south of Italy. Eng. Geol. **292** (2021). https://doi.org/10.1016/j.enggeo.2021.106270
20. Lazzarini, L., Antichi, M., Laboratori, S., Iuav, U.: Pietra d'istria: quarries, characterisation, deterioration of the stone of Venice. In: 12th International Congress on Deterioration and Conservation of Stone (2013)
21. Maravelaki-Kalaitzaki, P., Biscontin, G.: Origin, characteristics and morphology of weathering crusts on Istria stone in Venice. Atmos. Environ. **33**(11), 1699–1709 (1999). https://doi.org/10.1016/S1352-2310(98)00263-5
22. Yavuz, H., Demirdag, S., Caran, S.: Thermal effect on the physical properties of carbonate rocks. Int. J. Rock Mech. Min. Sci. **47**(1), 94–103 (2010). https://doi.org/10.1016/j.ijrmms.2009.09.014
23. Dionísio, A., Martinho, E., Pozo-António, J.S., Sequeira Braga, M.A., Mendes, M.: Evaluation of combined effects of real-fire and natural environment in a building granite. Constr. Build. Mater. **277** (2021). https://doi.org/10.1016/j.conbuildmat.2021.122327
24. Folk, R.L.: Practical petrographic classification of limestones. Am. Assoc. Pet. Geol. Bull. **43** (1959). https://doi.org/10.1306/0bda5c36-16bd-11d7-8645000102c1865d

25. Dunham, R.J.: Classification of Carbonate Rocks According to Depositional Textures (1962)
26. Molli, G., Conti, P., Giorgetti, G., Meccheri, M., Oesterling, N.: Microfabric study on the deformational and thermal history of the Alpi Apuane marbles (Carrara marbles), Italy. J. Struct. Geol. **22**(11–12), 1809–1825 (2000). https://doi.org/10.1016/S0191-8141(00)00086-9
27. Dedalostone Homepage. https://www.dedalostone.com/en/carrara-marble-types-and-colors/. Accessed 6 Sept 2021
28. Zendri, E., Biscontin, G., Bakolas, A., Moropoulou, A.: Scienza e Beni Culturali, Libreria Progetto, pp. 627–650 (1994)
29. Dean, N.E.: Geochemistry and archaeological geology of the Carrara Marble, Carrara, Italy. Class. Marble Geochem. Technol. Trade, 315–323 (1988). https://doi.org/10.1007/978-94-015-7795-3_34
30. Wielgosz-Rondolino, D., Antonelli, F., Bojanowski, M.J., Gładki, M., Göncüoğlu, M.C., Lazzarini, L.: Improved methodology for identification of Göktepe white marble and the understanding of its use: a comparison with Carrara marble. J. Archaeol. Sci. **113**(December), 2020 (2018). https://doi.org/10.1016/j.jas.2019.105059
31. Ingham, J.P.: Application of petrographic examination techniques to the assessment of fire-damaged concrete and masonry structures. Mater. Charact. **60**(7), 700–709 (2009). https://doi.org/10.1016/j.matchar.2008.11.003

Infrared, Terahertz and Air-Coupled Ultrasonic Non-invasive Inspection for Artworks: A Comparative Study on an Old Hand-Bound Book of the XIXth Century

Qinqian Lei[1,2], Hai Zhang[2(✉)], Stefano Sfarra[3], Ahmad Osman[4,5], Clemente Ibarra-Castanedo[2], and Xavier P. V. Maldague[2]

[1] School of Automation, Beijing Institute of Technology, Beijing 100081, China
[2] Computer Vision and Systems Laboratory (CVSL), Department of Electrical and Computer Engineering, Laval University, Quebec City G1V06, Canada
hai.zhang.1@ulaval.ca
[3] Department of Industrial and Information Engineering and Economics, University of L'Aquila, 67100 L'Aquila, Italy
[4] Department of Inspection of Components and Assemblies, Fraunhofer IZFP, 66123 Saarbruecken, Germany
[5] University of Applied Sciences (Htw Saar), 66117 Saarbruecken, Germany

Abstract. This work presents a comparative study for the inspection of an old book, which was bound in the XIXth century. Infrared thermography (IRT), terahertz time-domain spectroscopy (THz-TDS) and air-coupled ultrasonic C-scan (ACU) were used to retrieve the damages and defects in the book cover. Image processing techniques including Fast Fourier Transform (FFT), peak-to-peak and time-delay analysis in time domain were applied in order to highlight the shapes of subsurface features. X-ray radiography was used to anticipate the inspection results. Overall, this work appears useful to investigate the feasibility of different non-invasive inspection (NII) techniques applied to ancient books.

Keywords: NDT · Hand-bound book · Infrared · Terahertz · Ultrasound

1 Introduction

The use of non-invasive inspection (NII) techniques has magnificently been described in the recent literature [1]. Among the various applications that are possible with NII, its use into the cultural heritage (CH) field has attracted increasing attention, including architectural heritage [2], old paintings [3], books [4], and so on. In fact, the detection of subsurface defects such as splitting, cracks, tunnels caused by woodworms, foreign materials due to ancient restorations is a concern of any restorer [5]. Although an array of analytical techniques is useful in the cultural heritage field, the prevention is the best course of action with brittle materials. Thus, the use of NII techniques (that do not require collection of material) is of paramount importance [2].

A. Osman and A. Moropoulou (Eds.): *Advanced Nondestructive and Structural Techniques for Diagnosis, Redesign and Health Monitoring for the Preservation of Cultural Heritage*, SPM 16, pp. 115–126, 2022.
https://doi.org/10.1007/978-3-031-03795-5_5

In this work, the authors studied the advantages of various NII techniques for detecting the conservation status of an ancient book. Since ancient books may be damaged, defected and/or built with different materials, what is visible to the naked eye sometime doesn't exactly line up with the hidden reality. This explains why the use of NII techniques exploring different wavelengths (i.e., different depths in the sample under testing – SUT) is highly desirable [6].

In this paper, infrared thermography (IRT), terahertz time-domain spectroscopy (THz-TDS) and air-coupled ultrasonic C-scan (ACU) were used to retrieve damages in the book cover. Interestingly, additional defects were also detected because the SUT was handmade. Image processing techniques including Fast Fourier Transform (FFT), peak-to-peak and time-delay analysis in time domain were applied to the raw images to highlight the shapes of subsurface features. Finally, a comparative study of these techniques was conducted to summarize pros and cons.

2 Sample Under Test (SUT)

The SUT is an ancient book. It seems a children's book, published in Paris, France. The translated title (from French to English) shown on the book cover (Fig. 1) means «The little illustrated student». It belonged to Prof. Maldague's grandparents; the book was preserved in his parents' house. The purpose of the study is the detection of possible damages and defects, as well as the evaluation of sub-surface features by trying to understand the execution phase of the time. (The width of the book is 18 cm its height is 27 cm and its depth is 3 mm.)

Fig. 1. Photographs of the hand-bound book.

3 Methodology

In the following, the most important methods (Sects. "3.1 Infrared Thermography (IRT)", "3.2 Terahertz Time-domain Spectroscopy (THz-TDS)" and "3.3 Air-Coupled Ultrasound (ACU)") and techniques ("Fast Fourier Transformation") used in this work are explained for the sake of clarity of the reader.

The Discrete Fourier Transform (DFT) plays an important role in our case because it links the time domain to the frequency domain of the discrete signals, that is the most common one in physics compared to continuous signals. The Fast Fourier Transform (FFT) enhances the Discrete Fourier Transform (DFT) by making the transformation more practical and effective [7]. Here, FFT is used as a tool to analyze the IRT, THz and ACU images by obtaining the frequency spectrum of the detected signals.

The use of the FFT on thermographic data was first proposed by Maldague and Marinetti in 1996 [9], known as pulse phase thermography (PPT). PPT is able to retrieve defects by analyzing phase and amplitude data from a PT acquisition. Phase data is commonly analyzed owing to their tolerance both to nonuniform heating and environmental reflections [8]. More information regarding PPT can be found in [9]. THz-TDS measures the electric field and, therefore, provides information both on amplitude and phase of the THz wave through the use of FFT in frequency domain. Frequency information may be obtained from ACU images by using FFT, too.

3.1 Infrared Thermography (IRT)

Infrared thermography, one of the NII techniques has received increasing attention in the last few decades. Reasons for this include improved sensibility, higher resolution of thermal cameras and, also, advanced data processing technology. Several active thermography techniques have been proposed and, in this work, the pulsed thermography (PT) is used. In PT, the specimen surface is subjected to a heat pulse generated from a high-power heat source. Through a short pulse, the sample can be heated uniformly. As time goes, the surface cools down uniformly for a sample without any defects, while some subsurface discontinuities such as e.g., porosities, may chance the diffusion through the material by producing abnormal temperature patterns at the surface. The latter may be detected by an IR camera [10].

Figure 2 displays the experimental setup for pulsed thermography.

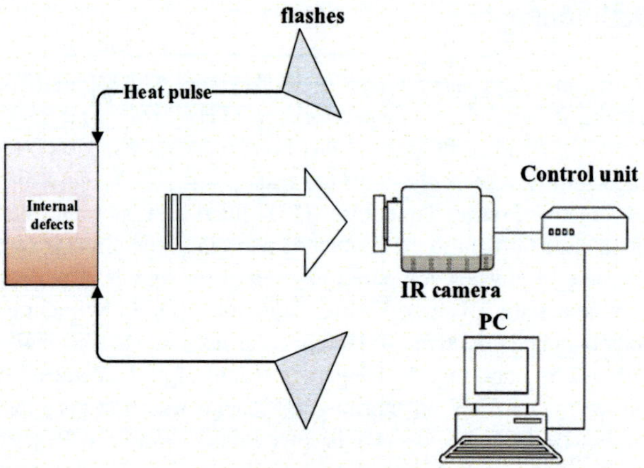

Fig. 2. Experimental set-up for PT

3.2 Terahertz Time-Domain Spectroscopy (THz-TDS)

The terahertz (THz) region of the electromagnetic spectrum lies in the gap between microwaves and infrared. THz has a series of ideal and unique properties, including the good transparency of many materials such as, e.g., cardboard. Also, a lot of materials can be identified in the terahertz range due to its distinctive characters exhibited [11]. However, the challenges it faced date back to the 1960s and the early 1970s were related to both a lack of reliable source and a sensitive detector [12]. The significant development came with the coherent terahertz generation and detection, known as THz-TDS. Its wide band measurement and the immunity to the background noise have attracted many researchers in a broad range of fields [13]. One more important thing is the two possible modes for THz images generation, i.e., the through-transmission and the pulse-echo modality. Because the inspected book cover was flat and not so thick, the through-transmission modality was considered the best candidate to detect possible subsurface defects [14].

Figure 3 shows the experimental set-up of the employed THz-TDS system. The pump wavelength was centered at 1550 nm, while the repetition rate was set at 100 MHz. The system had a 15 GHz frequency resolution. As already mentioned, the experiments were performed in transmission mode. The specimen was positioned in the focus of the optical system, between the THz emitter and the receiver [15].

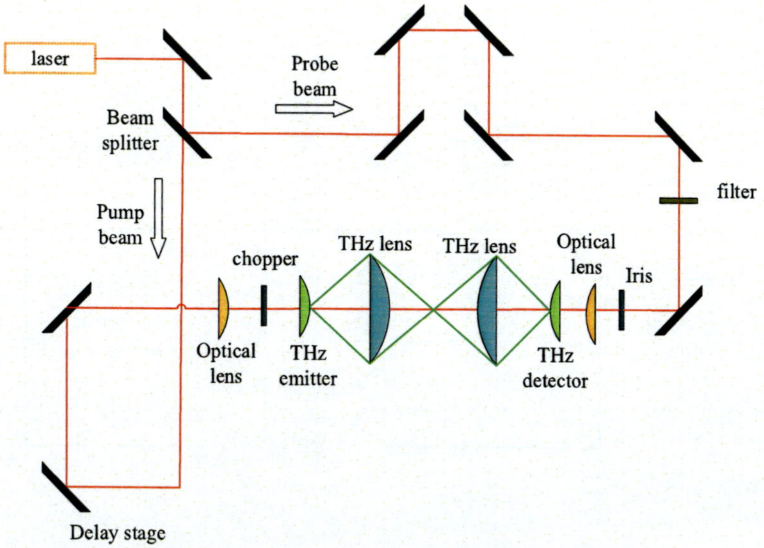

Fig. 3. Experimental set-up for terahertz time-domain spectroscopy.

3.3 Air-Coupled Ultrasound (ACU)

Classical ultrasonic C-scans are processed in water tank to obtain good coupling between the transducers and the tested samples, indicating contact and also a risk of water penetration in the specimen. A recent technique free the specimen from the link with water thanks to the use of new air-coupled transducers. This is an important achievement in NII, which even makes possible the performing of tests on non-waterproof materials [16].

The experimental set-up used for the air-coupled ultrasound (ACU) tests is shown in Fig. 4. Bearing in mind the size of the inspected object, also in this case the authors worked in transmission mode.

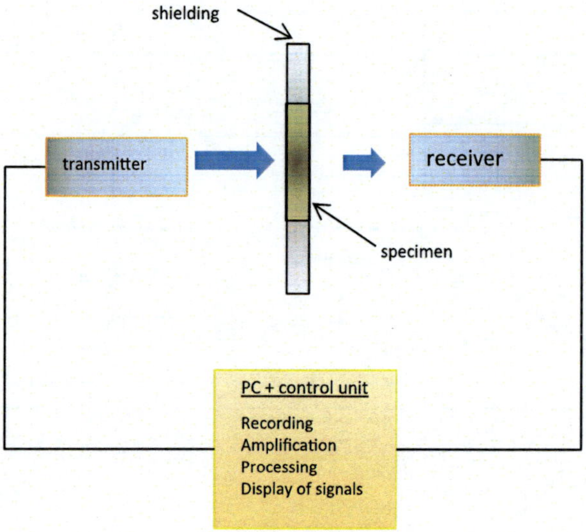

Fig. 4. Experimental set-up for ACU.

4 Result Analysis

The rear side book images is shown in Fig. 5. Reinforcement areas partially detectable in Fig. 1 (rear side – see the red dotted squares for reference) become clear in the phase result (Fig. 5). Additionally, the black dots (see the green dotted square on the right side of Fig. 5) can be related to tunnels caused by woodworms and not visible in the photograph. The worn area on the upper left-hand corner (see the red dotted square on the right side of Fig. 5) is more evident both in the amplitude and phase result.

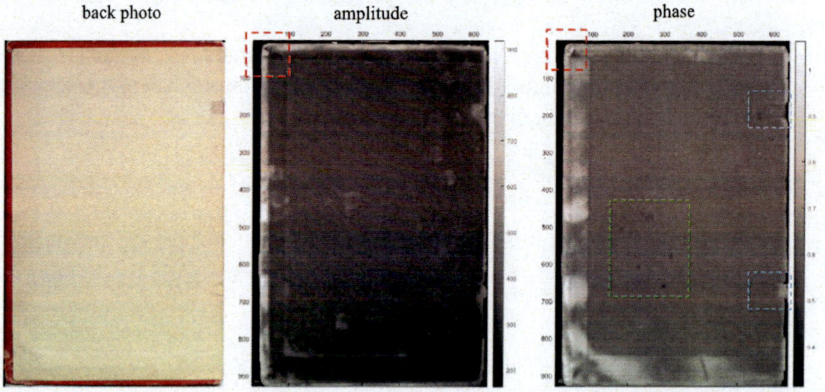

Fig. 5. IRT: Rear side of the book cover (on the left), amplitude (at the center) and phase result (on the right) at 0.1 Hz.

Figure 6 represents the result of the front side of the book cover acquired in frequency domain (0.1 Hz). Since foreign materials, such as, e.g., ashes or stains may limit the wavelength penetration, lower frequencies having higher anti-disturbance ability is analyzed; in such a case, signals may explore the book cover more deeply.

Fig. 6. IRT: front side photograph (on the left), amplitude (at the center) and phase (on the right) result at 0.1 Hz.

In Fig. 6, a series of black dots in the amplitude result (see the orange dotted rectangle) are evident. They can also be seen in Fig. 7; presumably, they are related to a variation of the heat conduction caused by defects due to woodworms.

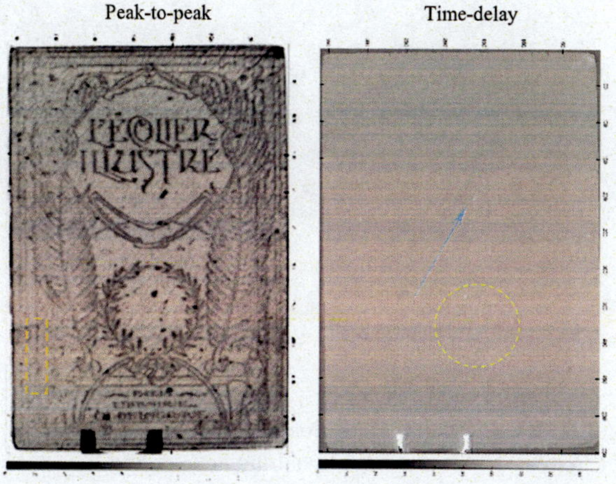

Fig. 7. THz-TDS: peak-to-peak (on the left) and time-delay (on the right) result.

Also, an irregular pattern (see the red dotted oval on the right side of Fig. 6) in the phase image resembles to a crack. As part of it, the whiter curved line (i.e., a part of the crack) may acts as starting point for a possible water ingress. Seeing the phase result of THz (Fig. 8(d)), this defect is also detectable (see the red dotted oval for reference) at the same position. Since this defect become evident only at higher frequencies, it should be located near the front side of the book cover.

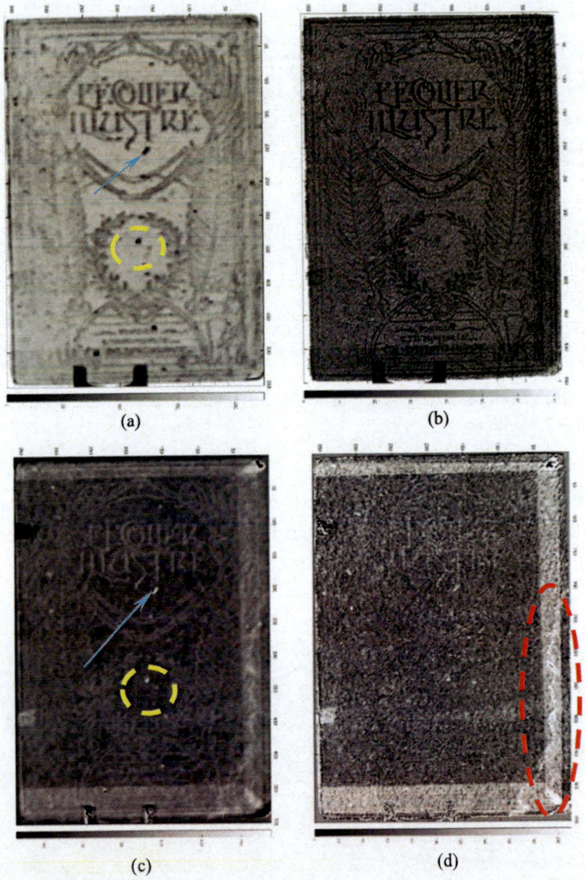

Fig. 8. THz-TDS frequency domain: amplitude results at (a) 0.28 THz and (b) 0.80 THz, and phase results at (c) 0.28 THz and (d) 0.79 THz

Moreover, there are two black dots in the phase result of Fig. 6; the first one located below the letter "T" (pointed out by a blue arrow), and the second one inside the laurel wreath (see the yellow dotted circle). The latter cannot be seen to the naked eye in the photograph of the front side of the book cover. They should probably foreign materials added accidentally during the manufacturing of the book.

However, these two signals also appear in Fig. 7 and Fig. 8. Figure 7 shows the THz-TDS inspection results. The time-domain peak-to-peak imaging reflects the absorption characteristic of the sample for whole THz radiation. Therefore, dark points in the peak-to-peak imaging can surely be linked to different absorption coefficients with respect to the surrounding area [17]. In the time-delay imaging, correlating the thickness and refractive index information of the sample, the dark points matching with the whiter areas. This indicates a different thickness due to the longer time delay.

In the amplitude image of THz technique (Fig. 8), the dark points appear more evident at lower frequencies, meaning that they are located deeper. Figure 8(c) and 8(d) show the phase information at 0.28 THz and 0.79 THz, respectively. The meaning is based on the refractive index at the corresponding frequency, by providing thickness information [17]. The two signals discussed above can be detected more clearly in Fig. 8(c) which has higher resolution.

In addition, in the time delay imaging (Fig. 7), the pattern on the book is obscure. This is caused by the limited depth resolution, which can be calculated as below:

$$\Delta L = c \times \Delta t \tag{1}$$

where ΔL is depth resolution, c is ultrasonic transmission speed, Δt is minimum time step.

Paper is considered as a random assembly of individual particles, namely fibers and pigments [18], thus it is hard to measure the propagation of light in paper. However, by estimation, the depth resolution is larger than the thickness of the pattern on the book cover.

The ACU results are shown in Fig. 9 and Fig. 10. time-delay can be used to determine the density of the materials, as below:

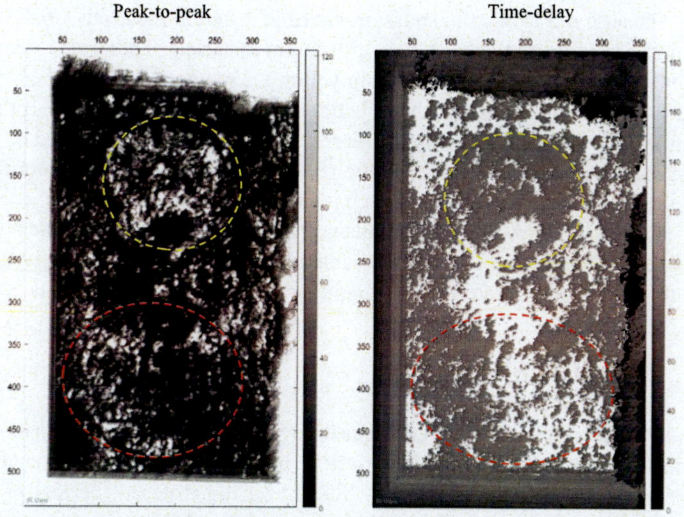

Fig. 9. ACU: peak-to-peak (on the left) and time-delay (on the right) results.

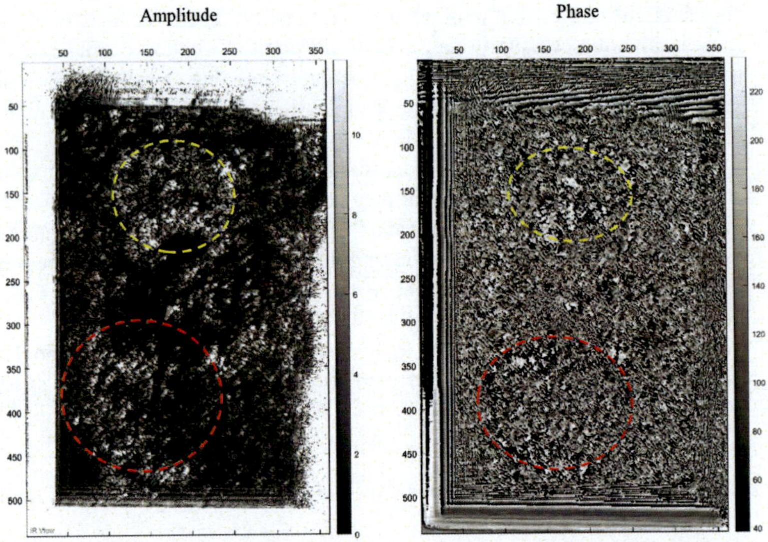

Fig. 10. ACU: amplitude (on the left) and phase (on the right) results.

$$\rho = Z/v \tag{2}$$

where ρ is material density, Z is acoustic impedance, v is sound velocity in the material.

Because ultrasonic transmission varies in different materials, so the scattered black points can indicate the existence of different materials. Therefore, it can be concluded that the title of the book (marked in yellow) is made of different materials from other parts of the book cover, which has better ultrasonic transmission. So is the laurel wreath in the darker region. Actually, the darker regions can also be caused by the increase of the material density, but refer to the peak-to-peak result, the brighter regions indicate higher absorption coefficients which is linked to different materials.

Figure 10 shows the ACU frequency domain results. The identical areas Fig. 9 are also detectable in Fig. 10, which are marked in yellow and red. However, the spatial and temporal resolutions are limited if compared to the THz-TDS results.

5 Conclusions

In this work, IRT and THz techniques provided more information than ACU. Both IRT and THz techniques were able to retrieve information concerning materials composing the book cover and its thickness. IRT shown the material thermal properties from a qualitative point-of-view. It should be noticed that THz technique can be used to analyze absorption coefficients related to materials as well as refractive indexes linked to thickness. Moreover, while IRT shown interesting details at high resolution (thanks to the thermal camera used), THz results displayed much information concerning the identification of defects. ACU in the TT mode is effective to identify the different materials and defects, but the spatial and temporal resolutions are limited.

FFT were used in all three methods by providing both amplitude and phase patterns. It is particularly efficient when combined with peak-to-peak and time-delay imaging obtained via THz and ACU technique by processing the data in time domain.

Acknowledgements. This research was supported by the Natural Sciences and Engineering Research Council (NSERC) Canada through the Discovery Grant program, Create-oN DuTy! Program, and the Mitacs Globalink Program.

This research was supported by the Canadian Foundation for Innovation (CFI)and the Canada Research Chair in Multipolar Infrared Vision (MiViM).

This research was also supported by the AutomaTiQ-Group of the htw saar.

The authors would like to thank Thomas Waschkies and Andrea Mross from Fraunhofer IZFP for their assistance in generating the ACU-data.

References

1. Shull, P.J.: Nondestructive Evaluation – Theory, Techniques, and Applications. CRC Press, USA (2002)
2. Moropoulou, A., et al.: Non-destructive techniques as a tool for the protection of built cultural heritage. Constr. Build. Mater. **48**, 1222–1239 (2013)
3. Abraham, E., et al.: Non-invasive investigation of art paintings by terahertz imaging. Appl. Phys. A **100**(3), 585–590 (2010)
4. Mercuri, F., et al.: Infrared thermography applied to the study of cultural heritage. Int. J. Thermophys. **36**(5–6), 1189–1194 (2014)
5. Sfarra, S., et al.: Discovering the defects in paintings using non-destructive testing (NDT) techniques and passing through measurements of deformation. J. Nondestruct. Eval. **33**, 358–383 (2014)
6. Tortora, M., et al.: Non-destructive and micro-invasive testing techniques for characterizing materials, structures and restoration problems in mural paintings. Appl. Surf. Sci. **387**, 971–985 (2016)
7. Nussbaumer H.J.: The fast fourier transform. In: Fast Fourier Transform and Convolution Algorithms. Springer Series in Information Sciences, vol. 2. Springer, Heidelberg (1982). https://doi.org/10.1007/978-3-642-81897-4_4
8. Duan Y., et al.: Quantitative evaluation of optical lock-in and pulsed thermography for aluminum foam material - ScienceDirect. Infrared Phys. Technol. **60**(5), 275–280 (2013)
9. Maldague, X., Marinetti, S.: Pulse phase infrared thermography. J. Appl. Phys. **79**, 2694 (1996). https://doi.org/10.1063/1.362662
10. Ibarra-Castanedo, C., Maldague, X.P.V.: Infrared thermography. In: Czichos, H. (ed.) Handbook of Technical Diagnostics. Springer, Heidelberg (2013). https://doi.org/10.1007/978-3-642-25850-3_10
11. Chan, W., Deibel, J., Mittleman, D.: Imaging with terahertz radiation. Rep. Prog. Phys. **70**(8), 1325–1379 (2007)
12. Gebbie, H.A., Stone, N., Findlay, F.D.: A stimulated emission source at 0.34 millimetre wave-length. Nature **202**(4933), 685–685 (1964)

13. Nishizawa, S., et al.: Terahertz time-domain spectroscopy (THz-TDS) approach to the quality control on pharmaceutical products. In: International Conference on Infrared Millimeter & Terahertz Waves, IEEE (2010)
14. Yuan, M., et al.: Numerical study on surface roughness measurement based on nonlinear ultrasonics in through-transmission and pulse-echo modes. Materials **14**(17), 4855 (2021). https://doi.org/10.3390/ma14174855
15. Zhang, H., et al.: An infrared-induced terahertz imaging modality for foreign object detection in a lightweight honeycomb composite structure. IEEE Trans. Ind. Inf. **14**(12), 5629–5636 (2018)
16. Imielinska, K., et al.: Air-coupled ultrasonic c-scan technique in impact response testing of carbon fibre and hybrid: glass, carbon and Kevlar/epoxy composites. J Mater Process Technol **157**, 513–522 (2004)
17. Zhang, Z., et al.: Terahertz time-domain spectroscopy for explosive imaging. Optik – Int. J. Light. Electron. Opt. **118**(7), 325–329 (2007)
18. Farnood, R.: Review: optical properties of paper: theory and practice. In: I'Anson, S.J. (ed.) Advances in Pulp and Paper Research, Oxford 2009, Transactions of the XIVth Fundamental Research Symposium Oxford, 2009, pp. 273–352, FRC, Manchester (2018). https://doi.org/10.15376/frc.2009.1.273

Nondestructive Assessment of the Resilience to Climate Change and Natural Hazards

A Multi- and Interdisciplinary Methodological Approach for Monitoring Cultural Heritage Built Assets: The HERACLES Experience

Amalia Siatou[1](✉), George Alexandrakis[2], Paraskevi Pouli[3], Antonella Curulli[4], Elissavet Kavoulaki[5], Snjezana Knezic[6,7], and Giuseppina Padeletti[4]

[1] Revive-art.works, Olofytou 1, 11142 Athens, Greece
amalia.siatou@gmail.com

[2] Institute of Applied and Computational Mathematics (IACM), Foundation for Research and Technology-Hellas (FORTH), N. Plastira 100, 70013 Heraklion, Greece

[3] Institute of Electronic Structure and Laser (IESL), Foundation for Research and Technology-Hellas (FORTH), N. Plastira 100, 70013 Heraklion, Greece

[4] Istituto Per Lo Studio Dei Materiali Nanostrutturati, Consiglio Nazionale Delle Ricerche (CNR), Rome, Italy

[5] Ephorate of Antiquities of Heraklion, 71202 Heraklion, Greece

[6] The International Emergency Management Society (TIEMS), Brussels, Belgium

[7] Faculty of Civil Engineering, Architecture and Geodesy, University of Split, Split, Croatia

Abstract. Current trends for the protection and safeguarding of Cultural Heritage (CH) support cautious and continuous documentation and monitoring of monuments as well as their surroundings with the aim to identify the origin of any alteration and take informed decisions before applying any mitigation action. To that end, a methodological approach and the related guidelines for a predictive and cost-effective monitoring of the alterations observed on a CH built asset are being proposed. This methodology considers the characteristic of the monument and its surroundings, with a thorough risk analysis, in a holistic approach which spans in different historic eras, materials and environments. The methodology was based on specific test-beds within the HERACLES (H2020-DRS-2015, GA700395) project, focusing on Climatic Change (CC) related events. The test-beds presented visual damages affected by CC that can be generalized to many other CH assets all over Europe. The concept follows protocol creation, defined in the form of checklists, aiming to estimate and classify the assessed alteration, in correlation with the factors affecting them. The overall goal is to incorporate the proposed methodology into everyday practice which will enable the creation of monitoring reports according to the needs and specificity of each monument/structure/material, supporting the total management cycle, ranging from the initial evaluation of the condition of the monument to specific mitigation actions. This methodology supported by scientific data, enables stakeholders or CH managers to prioritize mitigation actions based on informed decisions.

Keywords: Condition assessment · Monitoring · Built cultural heritage · Risk assessment

A. Osman and A. Moropoulou (Eds.): *Advanced Nondestructive and Structural Techniques for Diagnosis, Redesign and Health Monitoring for the Preservation of Cultural Heritage*, SPM 16, pp. 129–145, 2022.
https://doi.org/10.1007/978-3-031-03795-5_9

1 Introduction

Historical significance, materials, technological features, past treatments, and surrounding environment are important factors to be evaluated for the decision-making for safeguarding Cultural Heritage (CH). These factors together with a thorough risk analysis can conclude to a holistic approach on for the protection measures to be applied. Specifically, for CH built assets, risk analysis investigates how a factor (environmental, socio-economical, or other) can affect the value and integrity of the CH built asset, through different forms of alteration. For this reason, the proposed concept for the protection and safeguarding of monuments is also based on understanding and monitoring the area surrounding the monument, with the aim to clearly identify the origin and extent of alterations before any decision-making for their long-term protection.

In this framework, Pedersoli J. *et al.* [1], in their guide on risk management, presented a general methodology based on the different phases/steps strictly interrelated/interconnected, such as the analysis of the interaction between the monument or object with its environment (establishing the context), identification, analysis, evaluation, treatment of risks, and monitoring for an improved risk management. These steps are repeated during the lifetime of the monument. For analysing the risks, a tool has been created for calculating, comparing, and communicating the magnitude of the risk affecting CH. This tool consists of scales that are used to rate the magnitude of risk in terms of frequency, or rate of occurrence, and the expected loss of value for the investigated object [1].

The large-scale joint visualization of the spatial distribution of deterioration and environmental data is a common approach to present the risk to CH built assets. With the increasing threat from Climate Change (CC), it is important to focus on the impact of the environment on CH built asset, especially on buildings and sites, to provide solutions for mitigating CC impact. In fact, a reliable and accurate analysis of the interaction between the environment and CH asset is the foundation of the definition of critical decisions for preventive conservation and long-term protection.

Qian K. *et al.* [2] have proposed a visual analytical methodology, through a spatial view and distribution of deterioration risks, for the risk analysis of CH monuments; in particular, tangible parts of the monuments are considered by analyzing the physical damage of the monument. Through their study, they have created a correlation matrix representing the link between the environmental factors and the observed alteration. In this way, they have enabled a spatial "view" of deterioration risk to discover/identify distribution patterns and proposed a visual methodology for risk analysis [2]. Kioussi A. *et al.* [3] addressed the problem of creating risk indicators based on the correlation between the pathology of the monument, the external factors, and the physicochemical characteristics of the material. Their methodology, for qualitative and quantitative analysis, is based on the integrated documentation of all external and internal factors affecting the monument. Based on the study of the microstructure of building materials and their decay, they have identified several risks indicators which can be used for the decision-making process of a monument's preventive conservation and restoration [3].

The present work aims at describing the methodology developed and proposed within the HERACLES project [4], considering context and risk analysis, in a holistic and multidisciplinary approach, including all factors affecting a CH built asset, as well as previous works [1–3, 5] and the specific HERACLES test-beds experience [6–17]. Through this process the appropriate solution and procedures, for maintenance, conservation and restoration of CH built assets can be proposed [18–20]. This methodology can be used as a baseline of systematic documentation and monitoring approach that can be applicable in any situation. It correlates the condition of the CH built asset under examination with environmental, climatic, and territorial characteristics (context analysis) through risk assessment procedures and monitoring, thus covering a wider spectrum of required analysis on a CH built asset [18–21].

2 Methodology

The influencing effect of environmental, socioeconomic and historical factors on a CH built asset can be evidenced and evaluated in different ways. Our approach is based on the fact that only through systematic monitoring of all the factors affecting a CH built asset and recording the observed changes specific risks can be identified, and appropriate protection measures can be timely and effectively proposed. This methodology is based on the identification of external factors that are related to the observed alterations on the asset's condition. The condition is then documented either for the whole CH built asset or for specific Regions of Interest (RoI) (e.g., monument parts such as walls, wall paintings, rooms, columns) or Materials of Interest (MoI) (e.g., stones, mortars, metallic elements etc.). Based on the need to mitigate the factors negatively influencing the asset and to ensure its long-term preservation, actions can be proposed for the appropriate conservation and restoration and systematic maintenance. This methodology can be applicable to any CH built asset influenced by its surroundings, and by the CC effects.

The proposed methodology is outlined in five steps (Fig. 1). For each step, a checklist is created, followed by a correlation matrix, with the final aim of identifying the actions necessary for the long-term preservation of an asset. This is a straightforward methodology, where the effects of alteration are cross-linked to the alteration factors, with the scope to determine the type of action, which is best suited for appropriate preventive measures or remedial actions. This methodology can be applied either to the entire asset or to the RoI/MoI. For benchmarking the methodology, the environmental, climatic, territorial, archeological, and structural characteristic of the test-beds within the HERACLES (H2020-DRS-2015, GA700395) project where used. The HERACLES test beds cover a large range of the characteristics that can be found in monuments worldwide, since they expand in different eras, building methodologies, geographical and climatic environmental, and socioeconomic contexts [6, 7, 10–17] but also different effects of Climate Change (CC).

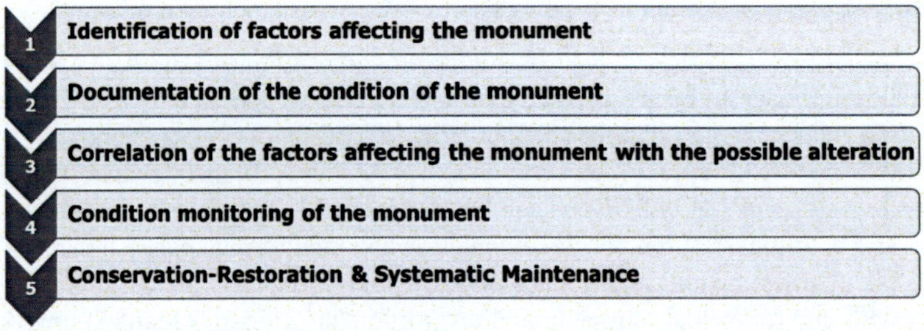

Fig. 1. The proposed methodology (modified from HERACLES deliverable D.7.2) [19]

Initially the methodology investigates, based on checklists, the external factors that interact with the monument/structure/material and documents their condition. The recorded alteration factors are then connected and correlated with the observed alteration. The information gained at this step is necessary to create reports, thus supporting, at the end of the whole process, the definition of the subsequent conservation-restoration or maintenance actions to be undertaken.

It needs to be noted that for the holistic approach, which includes different expertise, a common vocabulary was selected based on the CEN-TC 346 standards for general terms and alteration [9, 19, 21, 22].

More specifically the following steps are proposed:

2.1 Step 1: Identification of the Factors Affecting a Monument

The environmental, anthropogenic, geomorphological, and socioeconomic factors affecting a monument must be evaluated together with the specific alteration on its structure/material (Fig. 2). For example, the local climatic data provide information about the intensity and duration of rainfall, as well as the fluctuation of relative humidity and temperature or the intensity and direction of dominant winds. Additionally, through various sensors, the atmospheric pollution can be also recorded, while factors influencing/triggering corrosion can be identified. The systematic macroscopic observation plays a crucial role in detecting and monitoring any visible change. In the identification phase, a description of the external factors interacting with the monument is achieved through a checklist. Also, an evaluation of their impact can be generally assessed based on expertise and following criteria of relevance in each case-study. The thresholds in this assessment are ranging from "not applicable (0)" to "very high (5)" accounting for the influence of the individual factors (environmental/societal, structural, etc.) (Table 1).

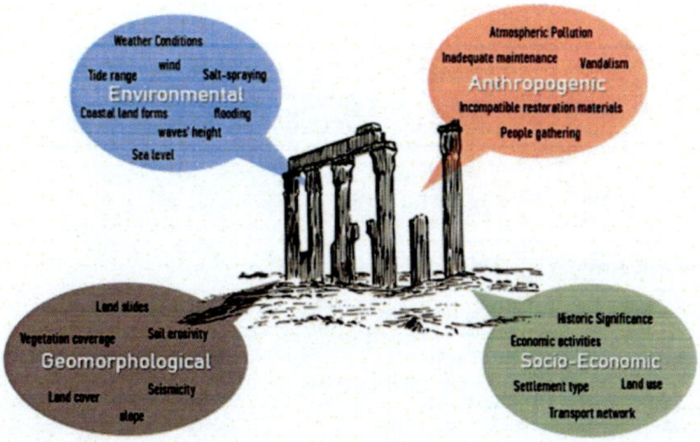

Fig. 2. Schematic representation of factors influencing a CH built asset.

Table 1. Example of checklist for evaluating external factors [19]

Factor	Value					
	n/a 0	Very low 1	Low 2	Moderate 3	High 4	Very high 5
Environmental						
Rain fall (mm/year)	n/a					
Atmospheric particles (aerosols)	n/a					
Freeze duration	n/a					
Freeze-Thaw cycles	n/a					
Temperature fluctuation	n/a					
Relative Humidity fluctuation	n/a					
Stagnant water accumulation	n/a					
Flooding	n/a					
Wind Aspect	n/a					
Fire Aspect	n/a					
Geomorphological						
Seismicity (g)	n/a					
Soil erosivity (mm/m^2)	n/a					
Landslide type	n/a					

(*continued*)

Table 1. (*continued*)

Factor	Value					
	n/a 0	Very low 1	Low 2	Moderate 3	High 4	Very high 5
Land cover	n/a					
Slope (O)	n/a					
Vegetation coverage (%)	n/a					
Coastal environment						
Coastal landforms	n/a					
Shoreline Changes (m/yr)	n/a					
Coastal Slope (%)	n/a					
Relative Sea-Level Rise (mm/a)	n/a					
Mean Wave Height (m)	n/a					
Mean Tide Range (m)	n/a					
Salt spraying	n/a					
Anthropogenic						
Atmospheric pollution	n/a					
Use of incompatible materials (m^2/total surface) (%)	n/a					
People gathering/pressure	n/a					
Inadequate maintenance	n/a					
Vandalism	n/a					
Socioeconomic						
Settlement	n/a					
Historical Significance	n/a					
Transport Network	n/a					
Land Use	n/a					
Economic activities	n/a					
Others						
Pests presence	Absent					

2.2 Step 2: Documentation of the Condition of the Monument

A qualitative, and to a certain extent quantitative, evaluation of the condition of the monument/structure/material can be achieved through visual inspection and observation techniques such as remote and in-situ sensing (e.g. by means of multispectral imaging [7]), optical macroscopy (e.g., endoscopic examination of openings in the walls) as well as microscopic methods (e.g., stereoscopic examination of the formation of crusts on the surface of a monument). In this case the check list is followed by a visual record. The pathology affecting the monument is assessed, similarly to step 1, in a scale ranging from not applicable (0) to very high (5), based on a qualitative macroscopic observation and evaluation of the damage or the decay extent, by experts and analytical data.

This can give important information on the intensity of the alteration phenomena and their impact to the CH built asset. This evaluation stage is mainly focused on the occurrence of the observed deteriorations/damages affecting the structural stability and/or the surface decay of the monument (Table 2). In Table 2 the percentages given for each value are indicatory and can be adjusted according to the specific monument/structure/material. This check list can be filled at regular time intervals leading to monitoring of the alteration phenomena on a monument/structure/material, as is explained in detail in step 4.

When necessary, based on results of the condition documentation, further specific diagnostics/analysis may be activated, such as diagnostics for assessing the structural stability of the whole monument (or specific structures) or physicochemical analysis of the deterioration of the surface.

Table 2. Example of a checklist for condition assessment based on macroscopic observation of the surface of the built CH asset [19]

Macroscopic observation of the condition of the monument							
Observed alteration		Value					
		n/a 0	Very low 1	Low 2	Moderate 3	High 4	Very high 5
Structural Damage							
General	Loss of connection/toothing between vertical elements (% crack amplitude/wall thickness)	n/a	<0.1	0.1–0.5	0.5–1.0	1.0–2.0	>2.0
	Loss of connection/toothing between horizontal and vertical elements (% crack amplitude/wall thickness)	n/a	<0.1	0.1–0.5	0.5–1.0	1.0–2.0	>2.0
Foundations and Vertical bearing elements	Differential settlements (% relative vertical displacement/horizontal distance)	n/a	<0.1	0.1–0.5	0.5–1.0	1.0–2.0	>2.0

(continued)

Table 2. (*continued*)

Macroscopic observation of the condition of the monument							
Observed alteration		Value					
		n/a 0	Very low 1	Low 2	Moderate 3	High 4	Very high 5
	Deterioration of binding media (mortar, etc.) (%/block thickness)	n/a	<10	10–20	20–30	30–50	>50
	Deterioration of blocks (stones and/or bricks) (%/element surface)	n/a	<10	10–20	20–30	30–50	>50
	Crack in the element (wall, pillar, column etc.) surface (%/element thickness))	n/a	<0.1	0.1–0.5	0.5–1.0	1.0–2.0	>2.0
	Crack in the element (wall, pillar, column etc.) thickness (spalling effect) (% length/element height)	n/a	<10	10–20	20–30	30–50	>50
	Inclination (% horizontal displacement/height wall)	n/a	<0.1	0.1–0.5	0.5–1.0	1.0–2.0	>2.0
Horizontal bearing elements (floors and vaulted structures)	Deterioration of binding media (mortar, etc.) (%/block thickness)	n/a	<10	10–20	20–30	30–50	>50
	Deterioration of blocks (stones and/or bricks) (%/wall surface)	n/a	<10	10–20	20–30	30–50	>50
	Crack in the vault surface (%/wall thickness)	n/a	<0.1	0.1–0.5	0.5–1.0	1.0–2.0	>2.0
	Maximum vertical deflection of horizontal load bearing elements (% with respect to span length)	n/a	<0.1	0.1–0.2	0.2–0.4	0.4–1.0	>1.0
Surface Alteration							
Surface decrease	Missing parts (%/total surface)	n/a	<10	10–20	20–30	30–50	>50
	Loss of material (% cm^2/total surface area)	n/a	<10	10–20	20–30	30–50	>50
	Superficial Cracks (% cm^2/total surface area)	n/a	<10	10–20	20–30	30–50	>50

2.3 Step 3: Correlation Matrix of Factors Affecting the Monument and the Observed Alteration

Environmental factors can induce physicochemical, structural, or aesthetic alteration on a built CH asset, evidenced in different ways. The matrix correlating environmental factors with any observed alteration, described in Table 3, will estimate the impact of the surroundings on the monument, in terms of different type of alteration and will help to "visualize" the evolution of changes for each monument/structure/material. This matrix correlates the factors affecting the monument based on the thresholds given in Table 1 and their subsequent documented impact based on thresholds given in Table 2.

Table 3. Correlation matrix (alteration factors/observed alteration)

		Alteration Factor											
		Environmental			Geomorphological			Anthropogenic			Others		
		N_1	N_2	...	...	...	...	...	...	...	...	N_{n-1}	N_n
Type of Observed alteration	n_1												
	n_2												
	...												
	...												
	...												
	...												
	...												
	n_{v-1}												
	n_v												

These values are then incorporated in a correlation matrix. This matrix correlates the factors affecting the monument based on the thresholds given in Table 1 and their subsequent documented impact based on thresholds given in Table 2. The calculation of the vulnerability for a given hazard, is expressed by one or more variables for a particular situation and/or RoI/MoI of the CH monument. The matrix correlating environmental factors with observed alteration and their impact is described in Table 4. Then the vulnerability for a given hazard, which is expressed by one or more variables for a particular situation and/or RoI/MoI of the CH, is calculated. An overall vulnerability for a particular hazard may be calculated as a function of the assigned risk value of specific variables:

Table 4. The generic table for vulnerability assessment

Factor	Alteration impact (exposure)[a]				
	1	2	3	4	5
	Very low	Low	Moderate	High	Very high
v_1	$x <$ or $> r_{11}$	$r_{12L} < x < r_{12U}$	$r_{13L} < x < r_{13U}$	$r_{14L} < x < r_{14U}$	$x <$ or $> r_{15}$
..					
v_n	$x <$ or $> r_{n1}$	$r_{n2L} < x < r_{n2U}$	$r_{n3L} < x < r_{n3U}$	$r_{n4L} < x < r_{n4U}$	$x <$ or $> r_{n5}$

[a]rijLorU – a lower or upper bound of a category's range for the specific variable, $i = 1 \ldots n$, $j = 1 \ldots 5$.

$$V = f(v_1, v_2, \ldots, v_n) \quad (1)$$

Risk assessment is estimated by considering the risk as "the probability of harmful consequences, or expected losses, resulting from interaction between vulnerability and exposure" by following the UNISDR (2009) definition of Risk the European Commission (ISO/EC, 2009).

Therefore, the vulnerability is estimated as the probability of risk, with the alteration impact functioning as the exposure variable. The factors are then multiplied by the values for observed alterations of the monument/structure/material. More specifically Risk was estimated by the equation:

$$R_i = V_i \times E_i \quad (2)$$

where V: Vulnerability and E: exposure variable.

These values are then incorporated in a correlation matrix where the vulnerability from the factors affecting the monument is multiplied by the values for observed alterations of the monument/structure/material. For example, environmental factors can induce physicochemical, structural or aesthetic alteration on a monument, evidenced in different ways. The matrix correlating environmental factors with observed alteration, described in Table 5, estimates the impact of the surroundings on the monument, in terms of different type of alteration and helps "visualize" the evolution of changes for each monument/structure/material. The values of the matrix range from a zero (0) to twenty-five (25) scale.

Table 5. Risk assessment graph for correlation matrix variables, modified from [18]

			Exposure of CH monument				
Vulnerability		**n/a 0**	**Very low 1**	**Low 2**	**Moder-ate 3**	**High 4**	**Very High 5**
EXCELLENT	1	0	1	2	3	4	5
VERY GOOD	2	0	2	4	6	8	10
GOOD	3	0	3	6	9	12	15
BAD	4	0	4	8	12	15	20
VERY BAD	5	0	5	10	15	20	25

In the case of multi-hazard risk evaluation, the application of the analytic hierarchy process AHP multicriteria method is proposed [18]:

$$R_i = V_i \times E_i,\ i = 1, .., n \quad (3)$$

where n is the number of hazards.

An overall risk (OR) is defined as:

$$OR = \sum\nolimits_{i=1}^{i} R_i \cdot H_i \tag{4}$$

where H_i i = 1, .., n, is a hazard.

The threshold values for the decision whether to make an intervention, are strictly dependent on the characteristics of the monument; therefore, the site manager/committee, also based on the priorities determined by the management authorities, should set these thresholds. Categories for the need of interventions can be described as:

- Very High, where immediate measures need to be implemented (15–25);
- High, where measures are of high priority (12–15);
- Significant, where some measures need to be implemented (8–12);
- Considerable, where measures need to be planned in due time (4–8);
- Tolerable, where monument/structure/material status is at present considered OK, but a plan for mitigation measures need to be established (3–4);
- Insignificant, where any interventions are of low priority for the health of the monument (1–3);
- Non-existent, when there is no correlation between alteration factor and alteration phenomena (0)

The condition initially identified represents "Time Zero" and is useful for the subsequent evaluations. For a substantial monitoring of the monument/structure/material the Tables 3, 4, 5 should be filled systematically and their variations with respect to "Time Zero" should be recorded. The evaluation of these results will support the decision-making in undertaking any intervention.

The aim of the proposed procedure is to provide a user friendly, compact, and fast methodology. Experience on the application of the procedure will allow to improve the calibration of the specific values reported in the different check lists corresponding to steps 1–3, which should be presently regarded as first estimations. For the initial assessment of a monument/structure/material condition ("Time Zero") all data collected are recorded and form an initial report, identifying the expertise needed for further evaluation and monitoring actions.

2.4 Step 4: Monitoring the Condition of the Monument/Structure/Material

The established checklists, along with the associated graphs, data and photographs must be updated at regular intervals (e.g. seasonal, monthly or yearly), identified for each case-study, and definitely after "crisis events" such as earthquakes, floods, landslides etc. These checklists are then incorporated into seasonal diagrams, in order to evaluate changes through time, as well as to evaluate the condition stability or the evolution of the alterations and the effect of each alteration factor.

Thus, through cross-time report logging, a monitoring table can be built and data can be visualized through graphs (see simulated example), depicting the influence of the alteration factors (step 1) affecting the examined monument, the evolution of the observed alteration (step 2) and in accordance on how they are correlated (step 3).

This information can help in the decision-making of the necessary actions for the long-term preservation of the monument, providing a useful documentation of the monument's condition.

2.5 Step 5: Conservation-Restoration and Systematic Maintenance

Generally, the actions that can be performed for any asset can be divided in three main categories. These categories according to [21] are:

- Preventive Measures; defined as "measures and actions aimed at avoiding or minimizing future damage, deterioration and loss and, consequently, any invasive intervention."
- Remedial Conservation and Restoration; where remedial conservation is defined as "actions applied directly to an object to arrest deterioration and/or to limit damage" and restoration is defined as "actions applied to a stable or stabilized object aimed at facilitating its appreciation, understanding and/or use, while respecting its significance and the materials and techniques used."
- Reconstruction; re-establishment of an object to an inferred earlier form, using existing or replacement material.

The application of each category on a monument and the related sub-actions are schematically represented in Fig. 3. Reconstruction was out of the scope of the project and therefore relevant actions were not further analysed.

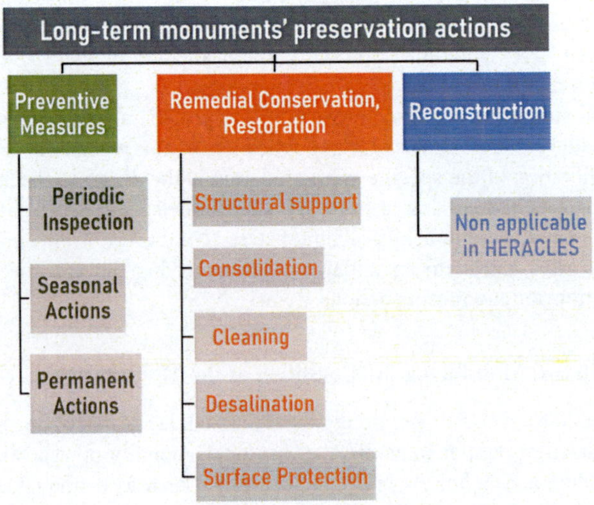

Fig. 3. Classification of actions for maintenance, conservation and restoration [19].

The actions described in Fig. 3 aim to prevent the effect of the environmental factors affecting the monument, mitigating or, if possible, eliminating any observed deterioration/damage, as well as improving its historic significance and aesthetic values.

In principle, preventive measures can be applied mostly to reduce the long-term effects of the factors affecting the monuments, whereas remedial actions and restoration are mostly applied in order to minimize the already occurred and observed damage or decay.

2.6 Simulation Example

To better explain the proposed methodology an example based on simulated values is presented. The environmental alteration factors (step 1) and the relating observed alterations on the monument (step 2) are monitored for 3 years at seasonal intervals (every 4 months) (step 4). Through the seasonal logging correlation matrices (step 3) and monitoring tables can be built.

In Fig. 4 the hypothetical behavior of the examined monument is defined through the monitoring table (4a) in regard to "Time Zero". The relevant graph (4b) helps in visualizing the trends of the external factors (red line) and the observed monument alteration (blue line). These trends are shown individually, and in terms of their correlation over time (green line), as well. The resulting trend is given by the variation from "Time Zero" (purple line), that provides an idea of the potential needs, since it considers the evolution of the correlated trends over time.

Monitoring a monument based on simulation data				
Month	Alteration Factor (0-5)	Observed alteration (0-5)	Correlation Matrix (0-25)	Variation from "Time Zero"
1	2	2	4	-
3	2	2	4	**0**
6	2	3	6	**+2**
9	3	3	9	**+5**
12	2	3	6	**+2**
15	2	3	6	**+2**
18	2	3	6	**+2**
21	3	3	9	**+5**
24	2	3	6	**+2**
27	2	3	6	**+2**
30	2	4	8	**+4**
33	3	4	12	**+8**
36	2	4	8	**+4**
40	2	4	8	**+4**

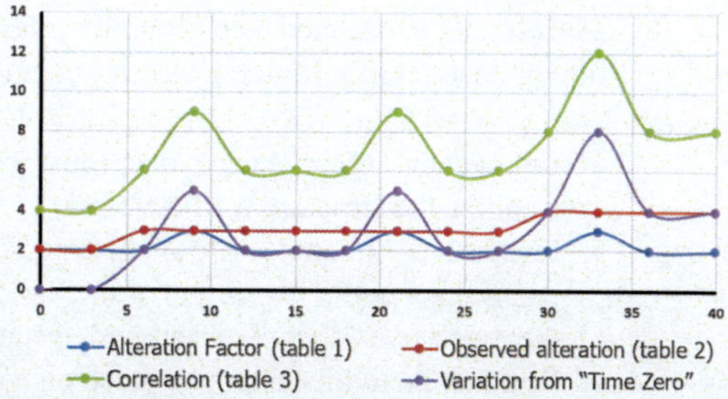

Fig. 4. Calculation of monitoring curves based on hypothetical data. Variation from "Time Zero" represents the general behavior of correlated factors over time [19].

2.7 Reporting and Monitoring of the CH Monuments

The possibility to set-up and implement an optimal and user-friendly overall approach, for understanding and correlating the external and internal factors affecting a monument, requires the availability of a comprehensive set of information. They can be derived from a series of reports collecting information on the condition of the monument/structure/material under study [4, 23]. They can be updated based on monitoring campaigns, which can be systematically planned. Thematic useful reports can also be proposed to assess the monuments' condition and help in the decision-making for planning of any mitigation action, these can be:

1. **Identification of the monument and its environment** (Historical survey, survey of the construction materials and techniques, environmental information)
2. **Condition Survey** (Survey of degradation of building materials, damage of building elements)
3. **Scientific diagnosis** (in-situ/field tests and analysis of samples material in the laboratory, assessment of the elements of the monument and its underground)
4. **Assessment of the structural safety** (Safety level of the building, modelling of the building, quantification of structural ability of the materials, of the acting loads)
5. **Conservation – Restoration plan** (architectural proposal, design of the rehabilitation works, conservation of the degradation of materials, repair and strengthening of the structural elements, upgrading of foundations, improvement of safety against crisis events such as earthquakes, landslides, extreme weather events, aesthetic amelioration, controlling surrounding environment, etc.).
6. **Preventive actions & Conservation-Restoration Treatments**; Based on the monument condition assessment, preventive or conservation restoration actions can be planned. The specificity of these actions is strictly dependent on the monument/asset characteristics.
7. **Monitoring**; Monitoring activities of different typology are important to assess the status of the asset and its surroundings, including environment. Moreover, the monitoring (continuous or periodic) can also be considered as a good practice to evaluate the evolution of the status of the asset and its surroundings.
8. **Maintenance**; Maintenance generally refers to actions to slow the rate of deterioration of an asset life. Maintenance is generally divided into three categories:

   - Emergency: Maintenance that must be carried out immediately in order to stabilize the structure for future use.
   - Preventive: Action taken to avoid expected failures. The simplest preventive maintenance is regular inspection of the building systems. This process also monitors the "service life" of materials and systems.
   - Routine: Activities that take place on a regular basis [22].

All the above-mentioned actions can be applied and adjusted according to the specific features and needs of the CH asset under investigation. These reports can assist in summarizing the needs of the monument and define the areas that need to be monitored. The same reports can be repeated/updated for monitoring the evolution of change or stability of the monument and the applied treatment/s.

3 Conclusions

This work focuses on the development of a general methodology for monitoring CH built assets based on the experience gained in the HERACLES project. The methodology implements check lists offering a best practice that can be applied to a wide range of built CH assets. It allows collecting and combining data regarding the surroundings (context) of a built asset and its condition. Then, a simple methodology for correlating the influencing factors and the observed alteration is proposed. It is based on average values, stemmed from literature and/or based on the experience and expertise of the different actors of the HERACLES project. This methodology establishes the initial values at "Time Zero", whereas regular and consecutive inspections will lead to the appropriate monitoring of a monument/structure/material by allowing the evaluation of the variations with respect to "Time-Zero". The condition of the monument/structure/material cross-time, combined with the use of analytical, sensing and monitoring techniques, is crucial also for understanding the effect that the different alteration factors can have on a monument/structure/material. Based on the previous analyses, the decision to intervene or not can be made. In addition, through regular monitoring and through the correlation of the observed parameters with the alteration factors/condition, trends associated with the CC may be identified. Once the decay mechanisms and their effects are understood, appropriate preventive measures can be suggested for limiting the negative impacts of the environment on the monument/structure/material. Additionally, remedial actions or restoration interventions can be better planned and monitored.

The data collected at "Time-Zero" can allow the creation/updating of appropriate reports and guidelines useful for the end-users. At "Time Zero" and successive statuses, the need of different expertise (such as civil engineers, specific scientific experts, conservator-restorers, architects etc.) is set for the further evaluation of the monument/structure/material. These experts will define and help end-users on how to fill in the checklists proposed. In this way, they can lead to the appropriate long-term maintenance/conservation of the monument/structure/material, through appropriate actions (monitoring, physicochemical analysis, focused actions).

The proposed methodology can be updated, optimized, improved, and refined, to create the appropriate correlation matrices that could apply in any case-study. This will help in better understanding the influence of the different alteration parameters regarding the observed behavior of the examined monument/structure/material. Through this procedure, thresholds can be defined to decide or not to make interventions and to prioritize treatment and/or interventions. This methodology will lead to a cost-effective maintenance of the CH asset effectively supporting decision-making process.

Acknowledgements. This work was supported by the European Union Frame-work Program for Research and Innovation HORIZON 2020 under Grant Agreement 700395 project HERACLES.

References

1. Pedersoli, J.L., Antomarchi, C., Michalski, S.: A guide to risk management of cultural heritage, ICCROM (2016). https://www.iccrom.org/wp-content/uploads/Guide-to-Risk-Managment_English.pdf. Accessed 12 June 2021
2. Qian, K., Sun, J., Chen, H., et al.: Visual analysis method for cultural heritage site risk assessment. J. Vis. **19**, 503–514 (2016). https://doi.org/10.1007/s12650-015-0325-7
3. Kioussi, A., Karoglou, M., Bakolas, A., Labropoulos, K., Moropoulou, A.: Documentation protocols to generate risk indicators regarding degradation processes for cultural heritage risk evaluation. Int. Arch. Photogramm. Remote Sens. Spatial Inf. Sci. **XL-5/W2**, 379–384 (2013). https://doi.org/10.5194/isprsarchives-XL-5-W2-379-2013

4. Padeletti, G.: Heritage resilience against climate events on site - HERACLES project: mission and vision. In: Moropoulou, A., Korres, M., Georgopoulos, A., Spyrakos, C., Mouzakis, C. (eds.) Transdisciplinary Multispectral Modeling and Cooperation for the Preservation of Cultural Heritage, pp. 360–375. Springer, Cham (2019). https://doi.org/10.1007/978-3-030-12957-6_26
5. Waltonm T.: Methods for monitoring the condition of historic places, Department of Conservation, Technical Series 27 (2003). https://www.doc.govt.nz/documents/science-and-technical/docts27.pdf. Accessed 12 June 2021
6. Alexandrakis, G., Kozyrakis, G.V., Kampanis, N.: Interventions on coastal monuments against climatic change. In: Moropoulou, A., Korres, M., Georgopoulos, A., Spyrakos, C., Mouzakis, C. (eds.) Transdisciplinary Multispectral Modeling and Cooperation for the Preservation of Cultural Heritage, pp. 385–401. Springer, Cham (2019). https://doi.org/10.1007/978-3-030-12957-6_28
7. Hatzigiannakis, K., et al.: Monitoring and mapping of deterioration products on cultural heritage monuments using imaging and laser spectroscopy. In: Moropoulou, A., Korres, M., Georgopoulos, A., Spyrakos, C., Mouzakis, C. (eds.) Transdisciplinary Multispectral Modeling and Cooperation for the Preservation of Cultural Heritage, pp. 419–429. Springer, Cham (2019). https://doi.org/10.1007/978-3-030-12960-6_29
8. Alexandrakis, G., De Vita, S., Di Vito, M.: Preliminary risk assessment ai Ustica based on indicators of natural and human processes. Ann. Geophys **62**(1) (2019). https://www.annalsofgeophysics.eu/index.php/annals/article/view/7765
9. CEN/TS 17135:2020: Conservation of cultural heritage - General terms for describing the alterations of objects, CEN-TC 346 Conservation of Cultural Heritage
10. Curulli, A., et al.: Methodology for the characterization of Cultural Heritage materials and their degradation state, using ex-situ laboratory analysis and microscopies: the HERACLES project approach. In: Geophysical Research Abstracts, vol. 20, EGU2018-19308 (2018)
11. Curulli, A., et al.: Multianalytical approach for the characterization of stone crusts. The case study of the Consoli Palace in Gubbio, Italy. In: Geophysical Research Abstracts, vol. 21, EGU2019-19077 (2019)
12. Curulli, A., et al.: A multi-disciplinary materials study as a contribution to evaluate degradation issues of monuments and archaeological sites towards. In: EGU2020-7010, Updated on 12 Jun 2020. https://doi.org/10.5194/egusphere-egu2020-7010
13. Veiga, J., et al.: Environmental influences on historical monuments: a multi-analytical characterization of degradation materials. In: EGU2020-21804 (2020). https://doi.org/10.5194/egusphere-egu2020-2184
14. Carvalho, F., et al.: The case study of the medieval town walls of Gubbio in Italy: first results on the characterization of mortars and binders. Heritage **1**, 468–478 (2018). https://doi.org/10.3390/heritage1020031
15. Catapano, F., Ludeno, G., Soldovieri, F., Tosti, F., Padeletti, G.: GPR surveys for soil and structural investigations at Gubbio town, Italy. In: 17th International Conference on Ground Penetrating Radar (GPR), Rapperswil, pp. 1–7 (2018). https://doi.org/10.1109/ICGPR.2018.8441544
16. Catapano, I., Ludeno, G., Soldovieri, F., Tosti, F., Padeletti, G.: Structural assessment via ground penetrating radar at the Consoli Palace of Gubbio (Italy). Remote Sens. **10**, 45 (2018). https://doi.org/10.3390/rs10010045
17. Curulli, A., Montesperelli, G., Ronca, S., et al.: A multidisciplinary approach to the mortars characterization from the Town Walls of Gubbio (Perugia, Italy). J. Therm. Anal. Calorim. **142**, 1721–1737 (2020). https://doi.org/10.1007/s10973-020-09937-9
18. HERACLES deliverable D1.3: Definition of methodologies for CC impact evaluation and risk and vulnerability analysis. https://cordis.europa.eu/project/id/700395/results. Accessed 12 June 2021

19. HERACLES deliverable D7.2: Methodologies for maintenance, conservation, and restoration. https://cordis.europa.eu/project/id/700395/results. Accessed 12 June 2021
20. HERACLES deliverable D7.1: Operational View Diagrams. https://cordis.europa.eu/project/id/700395/results. Accessed 12 June 2021
21. EN 15898:2019: Conservation of cultural heritage - Main general terms and definitions. Accessed 12 June 2021
22. HERACLES deliverable D1.1: Survey on guidelines and procedures for CH management. https://cordis.europa.eu/project/id/700395/results. Accessed 12 June 2021
23. Civil Protection: Scheda di 10 Livello di rivelamento danno, pronto intervento e agibilita per edifice ordinari nell'emergenza post-sismica, Protezione Civile, Conferenza delle regionie delle province autonome (2013)

On the Multi-hazard Risk Assessment of Urban Areas: Identification and Analysis of Exposure and Physical Vulnerability Indicators

Maria Xofi[1], José Carlos Domingues[1], Pedro P. Santos[2], Susana Pereira[2], Sérgio C. Oliveira[2], Eusébio Reis[2], José Luís Zêzere[2], Ricardo A. C. Garcia[2], Paulo B. Lourenço[1], and Tiago Miguel Ferreira[3(✉)]

[1] Institute for Sustainability and Innovation in Structural Engineering (ISISE), Department of Civil Engineering, University of Minho, Guimarães, Portugal
[2] Centre for Geographical Studies, Institute of Geography and Spatial Planning (IGOT), LA TERRA, University of Lisbon, Lisbon, Portugal
[3] Department of Geography and Environmental Management, University of the West of England, Bristol, UK
Tiago.Ferreira@uwe.ac.uk

Abstract. The development of integrated decision-support tools, able to assess multiple hazards at the regional and local scales, is a fundamental step to enhance the preparedness of urban areas to mitigate present and future risks arising from climate change. Understanding multi-hazard risk can help prioritize resilience-increasing actions and disaster prevention measures and form the basis for exploring institutional adjustments that improve stakeholders' capacities to manage risk. Within this framework, the present research work aims to identify and analyze a set of exposure and buildings' vulnerability indicators to be used as input to a multi-hazard risk assessment methodology. Exposure is to be measured using a dimensionless score resulting from the quantitative identification of the elements at risk. Regarding the physical vulnerability of buildings, it is to be evaluated on a hazard-by-hazard basis using a large-scale parametric-based vulnerability assessment approach. Finally, the obtained exposure and physical vulnerability indicators are to be put together in order to create different data layers, which are then used to identify hotspot risk areas. The Metropolitan Area of Lisbon, Portugal, is used as a pilot study area to discuss the applicability and potential of the proposed indicators.

Keywords: Urban risk · Vulnerability assessment · Exposed elements · Seismic hazard

1 Introduction

The potential impact of multi-hazard events on urban areas, i.e., events that include more than one natural hazard with interrelationships between the hazards that impact the same

A. Osman and A. Moropoulou (Eds.): *Advanced Nondestructive and Structural Techniques for Diagnosis, Redesign and Health Monitoring for the Preservation of Cultural Heritage*, SPM 16, pp. 146–155, 2022.
https://doi.org/10.1007/978-3-031-03795-5_12

location during the same time period, is, by definition, much more significant than the impact of a single hazard, making emergency response mechanisms unable to respond efficiently and effectively. However, although there has been a focus and a significant amount of research work on single hazard assessments, there is still limited research data for multi-hazard analysis. Hence, and given the increase in urbanization, it is essential to develop integrated risk assessment frameworks for measuring, managing, and mitigating the impacts of multiple natural hazards in urban areas.

Among the several natural hazards that in the past few decades have caused damages both to the building stock and infrastructure systems, as well as economic and human losses, earthquakes have been the most catastrophic ones.

The work reported in this paper aims to create a risk assessment framework for measuring impacts of multi-hazards in urban areas such as the Lisbon Metropolitan Area (LMA) by identifying, analyzing, and combining three core components: the seismic hazard, the exposure (i.e., built environment and population) and the seismic vulnerability of the exposed elements, to finally obtain the risk hotspots areas, addressing the linkage between the building typologies and the respective physical vulnerability. The validated seismic vulnerability results and the seismic hazard components are integrated into a Geographic Information System (GIS) tool developed in the open-source software QGIS to obtain the different seismic risk levels for the municipalities of LMA. Although the present work deals only with a single-hazard analysis (seismic), it establishes a simplified common methodology that can be applied to other natural hazards (such as floods and landslides) and will enable the creation of a framework for multi-hazard scenarios.

2 Methodological Framework: From Hazard, Exposure and Vulnerability Analysis to Risk Assessment

The methodological framework adopted in this paper for the risk assessment is based on the approach presented by Ferreira & Santos [1], including, besides a vulnerability and a hazard module, an exposure-related component.

The hazard level is obtained by combining three aspects: the maximum likely seismic intensity, PGA, and the geotechnical characteristics of the foundation soil. Considering a long-term time scale, it is the less dynamic component of the assessment.

The exposure is considered here based on the identification of the elements at risk, namely buildings and population. The various building typologies presented in the LMA are characterized, as is the number and the distribution of the population over the 3,015 km^2 that compose the LMA, see Subsect. 2.1.

The level of the physical vulnerability of the buildings is obtained through a parameter-based vulnerability assessment approach. From the evaluation of a few parameters of empirical nature, this approach allows obtaining a dimensionless index that measures the level of vulnerability of the building. Further details about the vulnerability assessment methodology can be found in Subsect. 2.2, and some representative vulnerability outputs obtained for the LMA are presented and commented in Sect. 4.

Seismic risk is then computed using the vulnerability-hazard matrix presented in Table 1, which combines the hazard and vulnerability results obtained using the above-described approaches.

Table 1. Seismic risk matrix.

Seismic Risk		Hazard: Low	Hazard: Moderate	Hazard: High	Hazard: Very High
Vulnerability	Low	Low	Low	Moderate	High
	Moderate	Low	Moderate	High	Very High
	High	Moderate	High	Very High	Extreme
	Very High	High	Very High	Extreme	Extreme

2.1 Identification of the Exposed Elements (Buildings and Population)

With 18 municipalities, 211 parishes, and near 2,813,000 inhabitants [2], the Lisbon Metropolitan Area (LMA) is the largest urban area in Portugal. According to the 2011 Census, the building stock in the LMA comprises 449,573 buildings spatially distributed by the 18 municipalities, as illustrated in Fig. 1. From these 449,573 buildings, 65% correspond to Reinforced Concrete (RC) structures and 34% to Unreinforced Masonry (URM) structures. The distribution of these two structural typologies within each one of the municipalities of the LMA is also provided in Fig. 1.

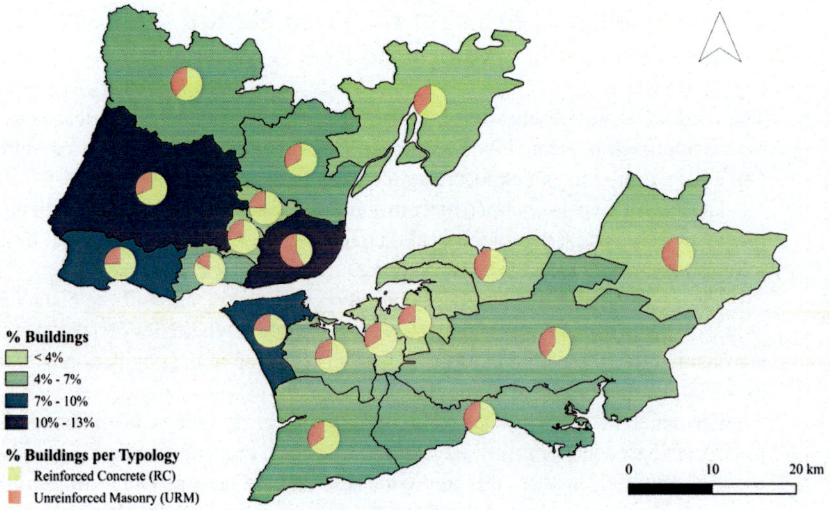

Fig. 1. Distribution of building stock and typologies per municipality.

Focusing specifically on the URM buildings, which is the typology addressed in this paper, four main types of unreinforced masonry buildings can be identified in the LMA [3, 4]:

- 'Pre-Pombalino' buildings – dating back to the period before the 1755 Lisbon Earthquake, these structures are characterized by irregular geometry, reduced dimensions in a plan, and narrow facades. With up to four stories, these buildings' walls are generally of poor-quality masonry (see Fig. 2, on the left).
- 'Pombalino' buildings – built in the aftermath of the 1755 Earthquake, the most distinctive feature of these buildings is the "Gaiola Pombalina," a timber-framed wall truss idealized to absorb the impact of horizontal seismic forces. These buildings typically present up to five stories, regular geometry, and large and regular in shape window openings (see Fig. 2, on the right).
- 'Gaioleiro' buildings – built between 1870 and 1930, these buildings can be seen as a downgrade compared to the previous "Pombalino" typology, with a lower level of construction quality (see Fig. 3, on the left).
- 'Placa' buildings – these report to a group of structures built during a very specific time period, mainly between 1930 and 1960, representing a structural solution characterized by a combination of masonry walls and reinforced concrete elements, such as concrete floor slabs (without any slab continuity). This is a transition structural typology between the traditional masonry structures and the modern reinforced concrete building construction (see Fig. 3, on the right).

Fig. 2. Unreinforced masonry buildings in the Lisbon Metropolitan Area: 'Pre-Pombalino' buildings (on the left) and 'Pombalino' buildings (on the right).

Fig. 3. Unreinforced masonry buildings in the Lisbon Metropolitan Area: 'Gaioleiro' buildings (on the left) and 'Placa' buildings (on the right).

Regarding the resident population in the LMA, its distribution is presented in Fig. 4. The distribution of the resident per building typology (URM and RC) is also provided in the figure. As can be observed in Fig. 4, Lisbon and Sintra are the two municipalities with the highest number of inhabitants with 548,358 and 377,823, respectively 19% and 13% of the total population of the LMA. When breaking down the population distribution per building typology, it is possible to observe that around 80% of the LMA's inhabitants live in RC buildings and 20% in URM buildings. In relative terms, Oeiras is the municipality that presents a higher percentage of RC buildings, about 83% of Oeiras building stock, whereas Lisbon is the one with a higher percentage of URM buildings, with near 57%.

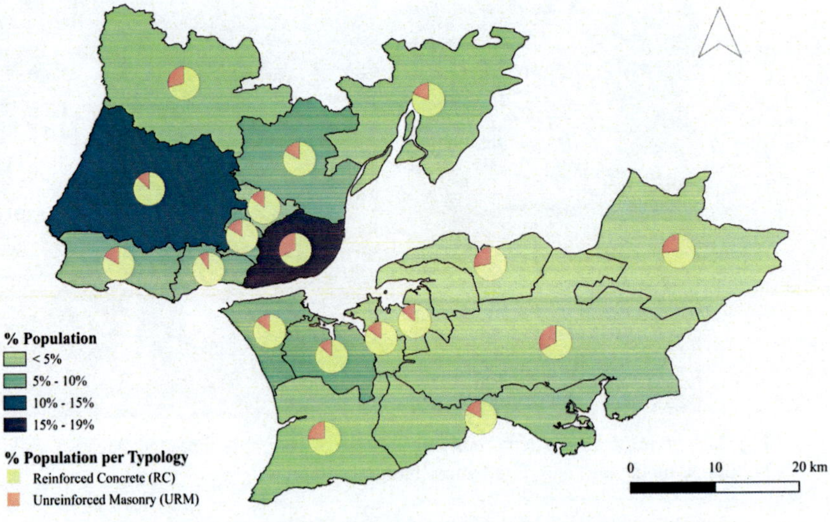

Fig. 4. Distribution of population per municipality and building typology.

2.2 Characterization of the Physical Vulnerability of the Building Stock

According to several authors [5–7], the selection of seismic vulnerability assessment methods should be based on three main essential criteria: level of detail, type of output (or scale of evaluation), and quality of the input data and tools (or methods) used. When dealing with a large number of buildings over a national or urban scale, the resources and quantity of information required are significant, and so the use of less sophisticated techniques or tools is more practical and necessary. Thus, methodologies for vulnerability assessments at a large scale should be based only on a few parameters, some of empirical nature.

The vulnerability index formulation applied in this work is based on the GNDT II level approach [8] for the vulnerability assessment of masonry buildings and is classified in the literature as a hybrid approach, combining the typological approach and the vulnerability index-based estimation. The seismic vulnerability methodology is based on post-earthquake damage observation and survey data covering several structural elements, which focuses on the most important aspects that define building damage, translated into a few parameters of empirical nature. It was originally proposed in Italy and has been applied over the last 30 years in many large-scale analyses. In 2011, it was adapted to the Portuguese masonry construction and improved further by introducing more detailed analysis for cases where adequate building data exist with new parameters related to the building's position and interaction between adjacent structures [9].

As in the original proposal, the methodology presented in this work can be used to obtain a seismic vulnerability index of buildings based on the evaluation of a few parameters of empirical nature. Each of these parameters corresponds to a specific feature that affects the seismic response of a building with a corresponding vulnerability class that is most applicable. These parameters are classified according to four vulnerability classes (C_{vi}) of A, B, C and D, and are associated with a weight (p_i) that defines the relative importance of each parameter to the overall seismic vulnerability of the building.

The current methodology was explicitly tailored to be fed by the Portuguese 2011 Census data survey, which subsequently constituted the basis for selecting the adopted parameters. Therefore, the following parameters in Table 2 were identified.

Table 2. Building parameters with vulnerability classes and weights.

Parameters		Classes C_{vi}				Weight
		A	B	C	D	p_i
P1	Structural system	–	5	20	50	2.5
P2	Period of construction	0	5	20	50	0.75
P3	Building position	0	5	20	50	1
P4	No. of storeys	0	5	20	50	0.5
P5	Ground plan layout	0	–	–	50	0.5

A total vulnerability index, I_v^*, is calculated using Eq. (1) by computing the weighted sum of the parameters multiplied by their specific weight assigned as a meaning of importance in terms of seismic response.

$$I_v^* = \sum_{i=1}^{5} C_{vi} \times p_i \tag{1}$$

For more straightforward interpretation and use, the vulnerability index, I_v^*, is then normalized to range between 0 and 100, assuming from that moment on the notation.

The parameters that mostly influence the seismic vulnerability of the URM buildings are parameter P1, which refers to the structural systems (including construction materials), and parameter P3 related to the relative position of the building within the aggregate. The parameter weights corresponding to the above mentioned are 2.5 and 1.0, respectively. Although less significant in terms of weights, the role of the other three parameters – P2 (Period of Construction), P4 (Number of stories), and P5 (Ground plan layout) – is also essential, contributing to capturing the overall seismic vulnerability of the building.

3 Analysis of the Physical Vulnerability Results

Based on the methodology described earlier in Subsect. 2.2, once all the seismic vulnerability indices per URM building have been computed, the results were spatially distributed using the GIS application software (QGIS 3.16.8-Hanover). Given the large scale of the Census building data, the physical vulnerability results were discretized per municipality, as displayed in Fig. 5.

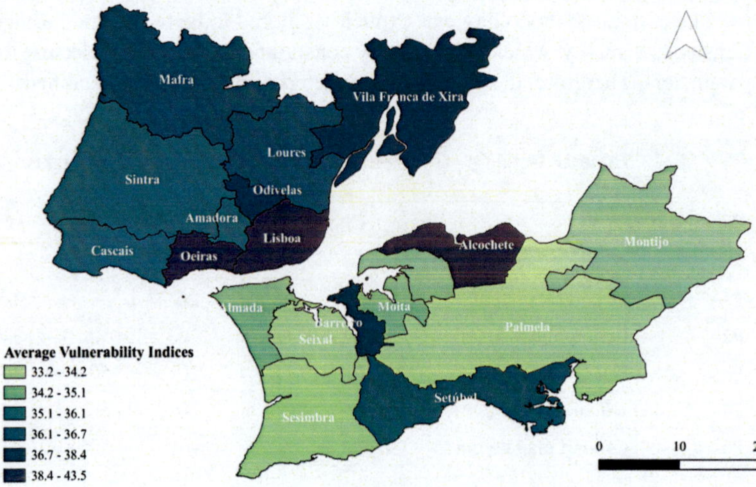

Fig. 5. Distribution of the average vulnerability indices obtained for the LMA municipalities.

Then, a representative selection of the most vulnerable municipalities related to the URM building typology was carried out based on the combination of the vulnerability outputs presented in Fig. 5 and the analysis of the portion of the buildings within the municipality whose vulnerability index value exceeds 50, given in Fig. 6.

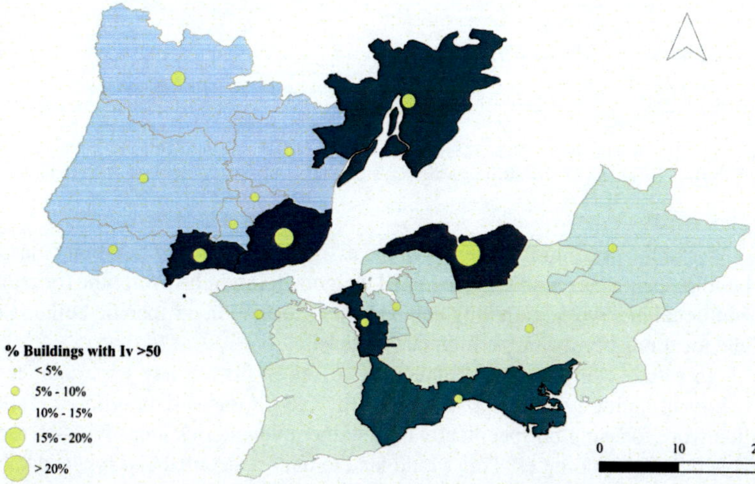

Fig. 6. Distribution of the percentage of URM buildings within the LMA municipalities with vulnerability index values higher than 50, with the identification of the six municipalities identified as the most vulnerable.

The six municipalities flagged in this first stage of analysis, namely Alcochete, Barreiro, Lisboa, Oeiras, Setúbal, and Vila Franca de Xira, should be therefore subjected to a more detailed assessment where all the factors that are contributing to this vulnerability must be carefully analyzed and understood. An example of this is provided in Fig. 7, where the spatial distribution of the physical vulnerability of Setubal municipality is displayed, highlighting with red color the areas in which the concentration of physical vulnerability is more significant.

As can be concluded from the observation of Fig. 7, most of the physical vulnerability hotspots identified in the municipality of Setubal are concentrated around Setubal city, a result that was already expected since this area concentrates most of the buildings. Despite that, it is possible to identify other urban areas where the level of physical vulnerability is also meaningful, such as São Sebastião and Brejos de Azeitão.

4 Final Remarks

The great advantage of this approach is the identification of the most vulnerable areas within the LMA, providing decision-makers and planners with valuable information for the higher risk areas, cities/towns, and even blocks or buildings (depending on the scale of the assessment carried out) on which retrofitting measures can be carried out.

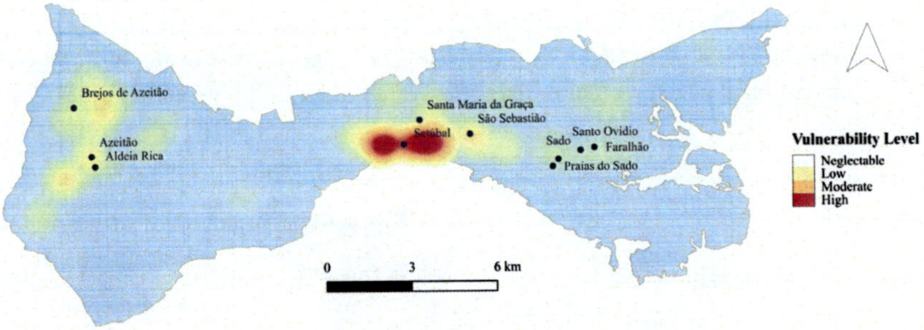

Fig. 7. Distribution of the physical vulnerability in Setúbal municipality.

As to the uncertainty inherent in the assignment of the vulnerability classes to each building parameter with the associated weights, it might be overcome (or reduced) with additional fieldwork, gathering detailed site information on specific building categories and locations of medium to high hazard levels.

In future work, it is recommended to refine further any uncertainties regarding unknown building characteristics from the 2011 Census data with carefully planned fieldwork focusing on specifically high vulnerable and risk areas with the most vulnerable building typologies. This would lead to a more detailed and robust database of the building vulnerabilities and possibly even to re-evaluate the parameters' weights.

The work presented in this paper is a part of a larger project aiming at developing an integrated risk assessment framework for measuring impacts of multiple natural hazards in urban areas, based on the comprehensive analysis of their direct and indirect interrelations and consequences. Such a framework is intended to constitute a useful decision-support tool, providing a singular standardized metric to measure aggregated urban risks and to accurately investigate the potential impact of pre and post-disaster strategies. Accordingly, this project will contribute to improving the protection of vulnerable communities to the impact of natural disasters.

The methodologies applied here for the seismic vulnerability assessment will also be developed and applied to assess flood, landslide, and fire vulnerability. Ultimately, this will allow for the creation of a multi-hazard framework, which can then be utilized by local emergency and risk mitigation planners to reduce the impact of any natural hazards to the most vulnerable communities.

Acknowledgements. The project "MIT-RSC - Multi-risk Interactions Towards Resilient and Sustainable Cities" (MIT-EXPL/CS/0018/2019) leading to this work is co-financed by the ERDF - European Regional Development Fund through the Operational Program for Competitiveness and Internationalization - COMPETE 2020, the North Portugal Regional Operational Program - NORTE 2020 and by the Portuguese Foundation for Science and Technology - GCT under the MIT Portugal Program at the 2019 PT call for Exploratory Proposals in "Sustainable Cities". Pedro Pinto Santos was financed through FCT I.P., under the program of 'Stimulus of Scientific Employment – Individual Support' within the contract CEECIND/00268/2017.

References

1. Ferreira, T.M., Santos, P.: An integrated approach for assessing flood risk in historic city centres. Water **12**, 1648 (2020)
2. Marques da Costa, E.: Socio-Economia. Atlas Digital da Área Metropolitana de Lisboa (2016)
3. Bernardo, V., Sousa, R., Candeias, P., Costa, A., Campos Costa, A.: Historic appraisal review and geometric characterization of old masonry buildings in Lisbon for seismic risk assessment. Int. J. Archit. Herit., 1–21 (2021)
4. Simões, A., Lopes, M., Bento, R., Gago, A.: Characterization of Lisbon old buildings. In: 15th World Conference on Earthquake Engineering (WCEE), Lisbon, Portugal, 24–28 September, pp. 14470–14480 (2012)
5. Vicente, R., et al.: Seismic vulnerability and risk assessment of historic masonry buildings. In: Structural Rehabilitation of Old Buildings, vol. 2, pp. 307–348. Springer, Heidelberg (2014)
6. Maio, R., Ferreira, T.M., Vicente, R.: A critical discussion on the earthquake risk mitigation of urban cultural heritage assets. Int. J. Disaster Risk Reduct. **27**, 239–247 (2018)
7. Ferreira, T.M., Mendes, N., Silva, R.: Multiscale seismic vulnerability assessment and retrofit of existing masonry buildings. Buildings **9**(4), 91 (2019)
8. GNDT-SSN: Scheda di esposizione e vulnerabilità e di rilevamento danni di primo livello e secondo livello (muratura e cemento armato), Gruppo Nazionale per la Difesa dai Terremoti, Roma, Italy (1994)
9. Vicente, R., Parodi, S., Lagomarsino, S., Varum, H., Silva, J.A.R.M.: Seismic vulnerability and risk assessment: case study of the historic city centre of Coimbra, Portugal. Bull. Earthq. Eng. **9**(4), 1067–1096 (2011)

Check for updates

Root-Cause Analysis on the Risk of Process Failure When Testing the Resilience of a Stone Masonry Building

Dimitra Passa[1], Stamatia Gavela[1(✉)], George Papadakos[2], and George Karydis[2]

[1] Department of Civil Engineering Educators, School of Pedagogical and Technological Education, 14121 Maroussi, Greece
matina.gkavela@gmail.com

[2] Domo+Lysis Lab - Laboratory of Structural Assessment and Protection, Athens, Greece

Abstract. The assessment of resilience on buildings that are recognized as members of cultural heritage is strongly based on laboratory testing methods. Root-Cause Analysis for incidents that occur, leading to the failure of the performed testing process, is required by the widely recognized International Standard ISO 17025:2017. This study presents the application of the Fault Tree Analysis (FTA) risk assessment technique for analyzing the root-causes of two indicative ways of failure that occur specifically when testing the resilience of stone masonry buildings: (a) the risk of sampling plan representativeness failure when performing mortar sampling and (b) the risk of false result (positive or negative) when investigating homogeneity on structural walls by implicating NDT methods (i.e. ultrasonic and GPR) in the process to detect the existence of voids, fissures and discontinuities. It is evident by the results of this study that the application of the FTA technique provides the basis of a precautionary measures selection and planning according to the actually leading parameters that eventually cause the failure of the testing procedure.

Keywords: Cultural heritage · Stone masonry · Root-cause analysis · FTA

1 Introduction

Building materials and structural members deteriorate with age and exposure to the weather. Historic masonry buildings have to show structural stability if they are used for any purpose. Moreover, they should be assessed whether they are vulnerable to be damaged under earthquakes or other physical conditions, for the building safety itself, as long as by the fact that they belong to cultural heritage. Then, after assessment of their condition, sometimes actions have to be taken for the restoration of the building. To ensure that those actions are proper, the assessment of the condition of the building and its materials should be based on reliable laboratory test results.

Both non-destructive and destructive techniques are used for this assessment. Non-destructive testing (NDT) techniques have been extensively used in the study of buildings

A. Osman and A. Moropoulou (Eds.): *Advanced Nondestructive and Structural Techniques for Diagnosis, Redesign and Health Monitoring for the Preservation of Cultural Heritage*, SPM 16, pp. 156–164, 2022.
https://doi.org/10.1007/978-3-031-03795-5_13

of cultural heritage because of the high cultural value of the concerned objects and the need to preserve them as intact as possible since sampling is not always possible or desired [1, 2]. NDT techniques can be combined with destructive tests, and this combination can provide information that NDT cannot provide [3]. One of the most employed NDT technique in the assessment of resilience on buildings, is the ultrasound method [4, 5]. Other NDT technique is ground penetrating radar (GPR), one of the most significant geophysical techniques applied in building inspection [6].

International Standard ISO/IEC 17,025 [7] enables laboratories to generate valid results. Especially when the laboratory is accredited according to this Standard, the validity of the results is enhanced. Two of the requirements of the Standard that ensure proper results in a masonry structure assessment process are:

a) The laboratory performing sampling is required to use a specific sampling plan. Especially when sampling for destructive testing has to be performed, a plan is required to ensure the reliability and representativeness of the testing result. This plan should be adequate, taking into account the value of the building which means repeating should be avoided. Moreover, sampling does not only refers to taking a sample but also to in situ non-destructive testing. In this case, also the laboratory is required to use a relevant plan.
b) The laboratory is required to plan actions to address risks and opportunities, and generally, corrective actions that are decided for the improvement of laboratory activity, all been referenced to the corresponding level of risk. Risk assessment on which actions to address risks and opportunities are based does not refer only to events that have already occurred, but also to events that are probable to occur. Risk assessment performed by a laboratory is a process required by the current version of the standard, instead of the required preventive actions that was included in the previous versions of standard ISO/IEC 17,025. Moreover, actions taken after an event, is required to be based on a root cause analysis of the event.

A commonly accepted technique for root cause analysis is Fault Tree Analysis (FTA), which is based on standard IEC 61,025 [8]. The technique can be used for the investigation of the root causes for an event that has already occurred (a real system failure). The same technique can be used preventively, for the risk assessment of laboratory activities (even before the occurrence of any event).

FTA is a technique primarily developed by airspace industry at early 60's [9]. Soon it was adopted by nuclear power industry. Currently, it is used in a wide variety of applications, i.e. industry systems assessment [10], investigation of accidents [11, 12], occupational health and safety management systems [13] and even for investigating failure mechanisms of measurement activities [14].

In the case of masonry structures, some unavoidable characteristic situations, such as the existence of inhomogeneous materials or age deterioration, exist. So root cause analysis for the determination of failure mechanisms must overcome those situations and go to further depth.

In this study, the application of FTA on the determination of root causes and for making decisions for relevant actions is presented. The aim is to assure that the right decisions are made for addressing preventive actions, within the context of laboratory activities improvement. Fault Tree Analysis (FTA) is presented in the case of root-cause analysis for two indicative ways of failure that occur specifically when testing the resilience of stone masonry buildings: (a) the risk of sampling plan representativeness failure when performing mortar sampling and (b) the risk of false result (positive or negative) when investigating homogeneity on structural walls by implicating NDT methods (i.e. ultrasonic and GPR) in the process to detect the existence of voids, fissures and discontinuities.

2 Method

As mentioned before, the technique that was used for performing the root-cause analysis on the risk of process failure when testing the resilience of a stone masonry building is the Fault Tree Analysis, as it is described in the IEC 61,025:2006. Table 1 provides a memo about the meaning of the used symbols. Commenting is also provided on the way each part of the diagram (an event or a gate symbol) should be dealt within the concept of the treatment for the corresponding testing process risks.

One of the most essential objectives of the FTA is to identify all the cut sets that lead to the top event. One advantage of the FTA diagrams is that they offer the opportunity to distinguish different cut sets within the mechanism that leads to the analyzed top event. According to IEC 61,025:2006, a cut set is "a group of events that, if all occur, would cause occurrence of the top event". Figure 1 presents a dummy FTA diagram where one of the cut sets is highlighted among a total of four cut sets in the diagram.

The probability of the top event, in general, is characteristic of the risk posed to the laboratory activities. Although, as it matters the management system of the laboratory, it is of most importance to identify which of the possible cut sets are characterized by significant probability of occurrence. Cut sets with the most significant probability will be the prevailing cases. When applying the ISO/IEC 17,025 requirements, the laboratory should consider treating the risks that correspond to the prevailing cut sets. This means that the laboratory should aim at applying actions to the primary events of the prevailing cut sets.

3 Analysis and Discussion

Two cases of risks that exist within the process of inspecting stone masonry buildings are presented. These examples represent the two major groups of testing processes that could be performed. The first corresponds to testing methods that require sampling and the second is an example of risks for in-situ techniques of inspection. In both cases, the overall risk posed by the fact that the inspected building is a non-homogeneous structure, so an actually representative inspection would be one with a maximized number of sampling points or points of in-situ measurements. Although, the fact that the building is a member of the cultural heritage, poses that the number of samples should be minimum in order not to harm the structure. Moreover, when non-destructive in-situ measurements are performed, the risk of not recognizing a part of the structure that has worn in a different grade than the rest of the structure is significant.

Table 1. Symbols for an FTA diagram according to IEC 61,025:2006 and corresponding impact to the treatment of risks related to the testing process.

Symbol	According to IEC 61025	Risk treatment opportunities
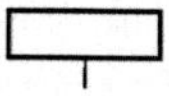	**top event** outcome of combinations of all input events	The unfavorable (or undesired) event that may occur as a risk for the performance of the laboratory which is developed through FTA in order to identify root-causes and apply efficient treatments
	gate symbol which is used to establish symbolic link between the output event and the corresponding events	
	gate **AND** the output event occurs only if all of the input events occur	Treatment of at least one of the input event is sufficient to cause improvement
	gate **OR** the output event occurs if any of the input events occur	All input events should be treated in order to achieve improvement
	**transfer** gate	Refers to another fault tree for further development
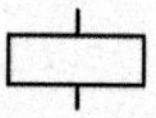	**intermediate event** event that is neither a top nor a primary event	Treatment of this event will not lead to significant reduction in the top events probability of occurrence as it is not a root-cause
	**basic event** event or state that cannot be further developed	Event that could be treated in order to manage the root-causes of the top event. This treatment could lead to significant reduction of the top event's probability of occurrence
	**house event** event which has happened or will happen with certainty	Treatment of this event or state is not feasible or is impossible
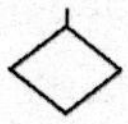	**undeveloped event** a primary event that represents a part of the system that is not yet developed	Treatment of this event or state is out of laboratory's concerns or existing information is not sufficient for further development. The laboratory is not able (at the current time) to treat this event or state

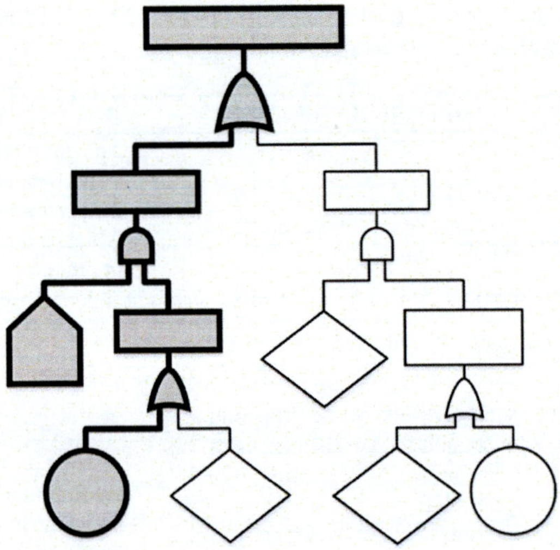

Fig. 1. Dummy FTA diagram presenting the meaning of a cut set.

Figure 2 presents an FTA diagram on the sampling plan representativeness failure. It is of great interest that two of the cut sets represented on the right part of the diagram depict that in order to avoid a lack of representativeness when performing the sampling plan, the laboratory should make a managerial effort on matters of duties, responsibilities and training among the laboratory personnel. In the same cut set, the fact that the inspected building is a worn structure is a very strong aspect of the applied inspection process, but also something that cannot be managed by the laboratory.

Other aspects of the sampling process could be managed, but on the basis of preparation that is out of the scope of the laboratory activities. Such a case would be a request for sampling on the stone masonry wall of a monastery in the "Meteora" complex which is on the side of the cliff. In such a case, preparation is sometimes possible, but it is out of the competence of the laboratory, as the person performing the sampling should be an expert on sampling and a climber at the same time. This is an existing risk for the laboratory's activity of sampling, but it is a risk that cannot be treated by a typical laboratory by itself.

Similar findings come from reviewing also the FTA diagram on the risk of negative or positive false result of the Non- Destructive Inspection (NDI) process when the homogeneity of the stone masonry wall is inspected. This process could be, in many cases, a preliminary step before implementing a sampling plan for retrieving specimens from the masonry structure.

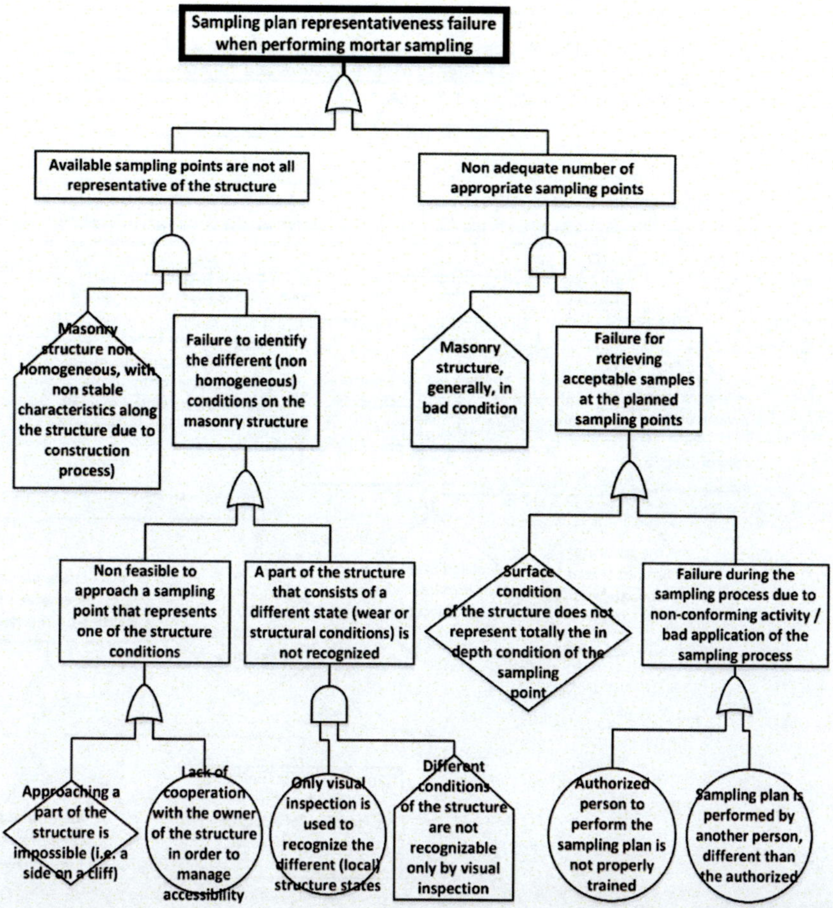

Fig. 2. FTA diagram analyzing the sampling plan representativeness failure mechanism when performing o mortar sampling on a stone masonry building.

As it is shown in Fig. 3, the fact that voids, fissures and discontinuities that are to be located and assessed are, generally, too small, is a situation that will always be certain in such inspection processes. Although, it is a fact that the laboratory cannot alter or avoid. Instead, even if standard references on the inspection process do not provide a clear decision rule on whether a stone masonry structure is homogeneous or not, the laboratory could proceed on building such a rule, taking into consideration the possible extent or dimensions of the inspected voids, fissures and discontinuities.

The aspect of appropriate management of duties assignment and training of person appears also in the case of NDIs. Also, apart from the risks of a bad application of the sampling plan, a risk of using a poorly described and analyzed sampling plan is another cause of NDI failure, but its root-causes are deeper. In this case, Fig. 3 refers to another FTA diagram, possibly a diagram like the one in Fig. 2.

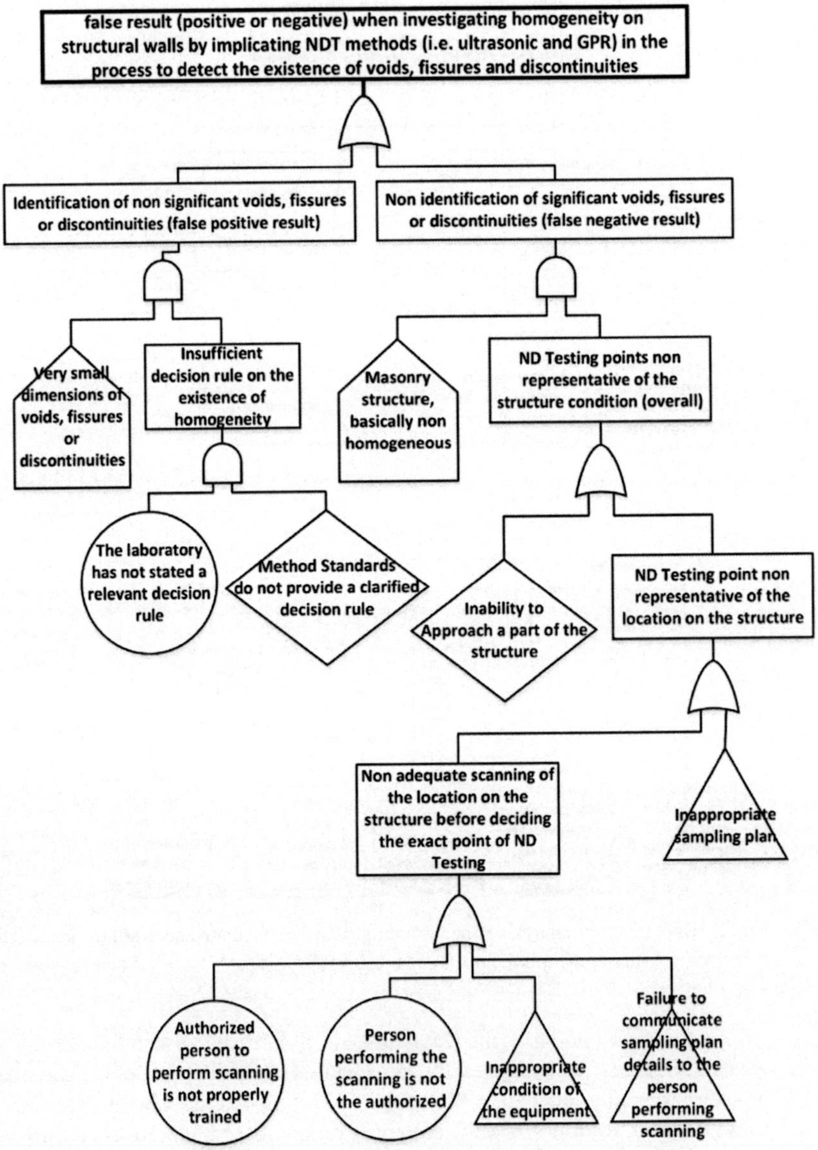

Fig. 3. FTA diagram analyzing the failure mechanism (positive or negative false result) when investigating homogeneity on stone masonry walls by implicating NDT methods.

An attempt for a tabular representation of the same context of risk analysis as in the two FTA diagrams would appear like in Table 2. The content of the right-most column of the table providing possible root-causes for the same risks is correct. This information could be retrieved easily during a brainstorming of a team of persons involved in such inspection processes. Although, the way of presentation of this information that is identical to parts of the FTA diagrams could lead to misinterpretations on which of these causes are actually root-causes. The group of experts would easily notify that there is a significant probability of sampling plan failure (in both cases), but maybe they would not proceed to a deeper analysis as they would feel satisfied of their analysis. In such a case, they would have noticed the problem, but probably they wouldn't solve it as a managerial problem of duties assignment and training scheduling.

Table 2. Tabular representation of the same root-cause analysis as by FTA diagrams in Fig. 1 and Fig. 2.

Context	Required aspect (criteria)	Adverse event	Root-causes
Sampling	The sampling plan assures the representativeness of sampling	Sampling is not representative of the inspective stone masonry structure	• Non-homogeneous structure • Parts of the structure in bad condition • Inability to approach parts of the structure • Failure during the sampling procedure (bad sampling)
In-situ ND Inspection	Check for homogeneity on the structural walls	Identification of non-significant or non-identification of significant voids, fissures or discontinuities	• Non-homogeneous structure • Very small dimensions of voids, fissures and discontinuities • Not appropriate sampling plan (procedure to identify the measurement points) • No decision rule on the existence of homogeneity

An FTA cannot provide all the information provided by a tabular risk assessment technique. A table created to serve within an FMEA or a HAZOP process may include a great number of columns analyzing different aspects of the risk assessment process. A combination of these two approaches (a tabular and a diagram like within an FTA) is practicable. Whenever a tabular analysis includes a part of analyzing the root causes of the risk and it seems that the corresponding ability of the tabular analysis (the content of one or two cells) is not adequate, then a reference to a related FTA diagram could be a proper solution.

4 Conclusions

A commonly accepted technique for root cause analysis is Fault Tree Analysis (FTA). The application of FTA risk assessment technique can be used for analyzing the root-causes of failure mechanisms that occur when testing the resilience of stone masonry buildings. FTA is an efficient tool for accredited laboratories in order to decide actions to address risks and opportunities. There is no explicit need to use FTA only for investigating events that have already occurred. Laboratories that perform inspections on stone masonry buildings may use FTA also as a precautionary technique to prevent undesired situations. If the characteristics of stone masonry buildings (lack of homogeneity or extensive age deterioration) are well analyzed and the primary events (the root causes) leading to undesired top events are identified, the corresponding risk is expected to be minimum.

References

1. Moropoulou, A., Labropoulos, K.C., Delegou, E., Karoglou, M., Bakolas, A.: Non-destructive techniques as a tool for the protection of built cultural heritage. Constr. Build. Mater. **48**, 1222–1239 (2013)
2. Kordatos, E.Z., Exarchos, D.A., Stavrakos, C., Moropoulou, A., Matikas, T.E.: Infrared thermographic inspection of murals and characterization of degradation in historic monuments. Constr. Build. Mater. **48**, 1261–1265 (2013)
3. Menéndez, B.: Non-destructive techniques applied to monumental stone conservation. In: Marquez, F.P.G. (ed.) Non-Destructive Testing (2016)
4. ASTM D 2845-00: Standard test method for laboratory determination of pulse velocities and ultrasonic elastic constants of rock
5. ASTM C 1383-15: Standard test method for measuring the p-wave speed and the thickness of concrete plates using the impact-echo method
6. ASTM D 6432-19: Standard guide for using the Surface Ground Penetrating Radar method for subsurface investigation
7. ISO/IEC 17025:2017: General requirements for the competence of testing and calibration laboratories
8. IEC 61025:2006: Fault tree analysis (FTA)
9. Ericson, C.: Fault tree analysis by design. In: Proceedings of the 16th International System Safety Conference (1998)
10. Faure, J.-M., Barragan, I.S., Papadopoulos, Y.: Including systematic faults into Fault Tree Analysis. SAFEPROCESS **06**, 811–816 (2006)
11. Kontogiannis, T., Leopoulos, V., Marmaras, N.: A comparison of accident analysis techniques for safety-critical man–machine systems. Int. J. Ind. Ergon. **25**(4), 327–347 (2000)
12. Nivolianitou, Z., Leopoulos, V., Konstantinidou, M.: Comparison of techniques for accident scenario analysis in hazardous systems. J. Loss Prev. Process Ind. **17**, 467–475 (2004)
13. Marhavilas, P., Koulouriotis, D., Mitrakas, C.: Fault and event tree techniques in occupational health – safety systems – Paer I: integrated risk evaluation scheme. Environ. Eng. Manag. J. **13**(8), 2097–2108 (2014)
14. Petrík, J., Blaško, P.: Fault tree analysis for Brinell hardness measurement. In: Proceedings of the Scientific Conference Quality and Leading Innovation (2014)

Digital Heritage, a Holistic Approach

The Role of UAV and Laser Scanners in the Post-earthquake Assessment of Heritage Buildings After the 2020 Earthquakes in Croatia

Mislav Stepinac(✉), Luka Lulić, and Karlo Ožić

Faculty of Civil Engineering, University of Zagreb, Zagreb, Croatia
mislav.stepinac@grad.unizg.hr

Abstract. In the wake of recent strong earthquakes in Croatia, there is a need for a detailed and more comprehensive post-earthquake damage assessment. Given that masonry structures are highly vulnerable to horizontal actions caused by earthquakes and a majority of the Croatian building stock is made of masonry, this field is particularly important for Croatia. In this paper, the "regular" post-earthquake assessment procedures of damaged buildings are given and new technologies such as photogrammetry, laser scanning and UAV are shown in order to preserve the built heritage. As Zagreb was severely damaged by the earthquakes in 2020, there is an urgent need to make, at least, digital twins of representative buildings. In this manuscript, different case studies with different demands for preservation are shown.

Keywords: Assessment · Earthquake · Zagreb · Laser scanning · Cultural heritage · UAV · Photogrammetry

1 Introduction

Two major earthquakes hit Croatia in 2020. The first in March with an epicenter about 10 km from Zagreb, medium magnitude ML = 5.5 from, and the second of medium magnitude ML = 6.3 with an epicenter about 50 km from Zagreb [1]. The first earthquake significantly damaged the city of Zagreb and its surroundings, and the second one, to which the epicenter was a little further, additionally damaged already damaged buildings.

The earthquakes damaged most of the buildings in the Lower Town and Upper Town – historical centres of Zagreb, including residential buildings, universities, schools, kindergartens, hospitals and public buildings. The larger part of the city's historical centre (Upper and Lower Town) was severely damaged because the buildings in the centre were built before any seismic regulations. The damage to historical buildings is enormous. Numerous museums, churches, university buildings have been severely damaged (Fig. 1).

As well as most parts of the European region, many existing buildings in Croatia are built in masonry. Given that most of the so-called "strategic" buildings of cultural significance and high historical importance are built using masonry, such a condition is

A. Osman and A. Moropoulou (Eds.): *Advanced Nondestructive and Structural Techniques for Diagnosis, Redesign and Health Monitoring for the Preservation of Cultural Heritage*, SPM 16, pp. 167–177, 2022.
https://doi.org/10.1007/978-3-031-03795-5_3

Fig. 1. Typical damage to cultural protected buildings after the Zagreb earthquake: damage to the residential buildings; damage of the church and the Zagreb Cathedral; damage of the dome of one of the buildings in the Lower Town (Photo credit: M. Stepinac)

suggesting that the assessment and rehabilitation of existing masonry structures must be conducted on a very high level [2–6]. Thus, the regular post-earthquake assessments should be followed by using state-of-the-art technologies for the preservation and digitalization of cultural heritage buildings. The construction sector is the slowest one to appropriate new technologies [7, 8] - and this is something that should be changed. This motivates the quest for finding the technological solution for the safety assessment of existing structures and the digitalization of cultural heritage buildings [4].

In this paper, the situation of heritage protected buildings after the recent earthquakes are shown. Furthermore, the post-earthquake assessment was briefly explained and digitalization of heritage properties is shown on four different case studies. Unfortunately, only four of the many case studies are shown due to paper size limitations. One of the case studies is described in detail in [9].

2 Position and Historical Background of Zagreb

Cities in continental Croatia were built on the legacy and principles of construction in Austro-Hungary from the beginning of the 20th century. Zagreb's historic urban complex is a protected area regulated by the Law on the Protection and Preservation of Cultural

Heritage. The area is divided into two zones, zone A and zone B (Fig. 2) [1]. Zone A includes the oldest and most architecturally valuable parts of Zagreb and is characterized by densely built blocks of buildings made of stone, brick or a combination of materials.

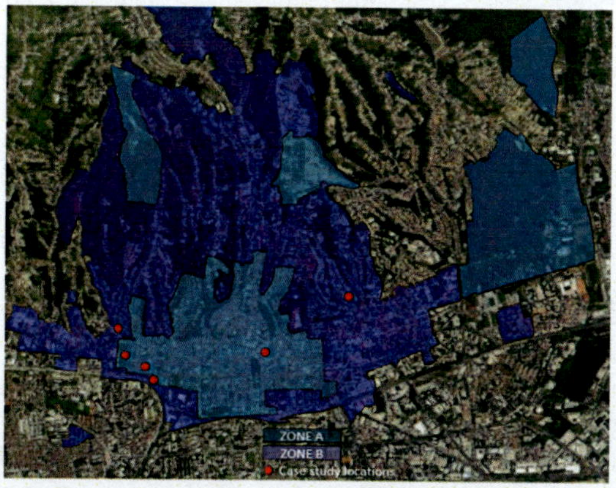

Fig. 2. Protected zones A and B with the location of the case studies shown later in the paper

The masonry building aggregates are not rare in Zagreb and other European cities. The behavior of buildings in the aggregate is extremely complex, where many parameters affect their response during an earthquake. Also, buildings at the ends of aggregates are often more severely damaged than those inside aggregates during earthquakes [10, 11]. Modeling of the entire aggregate or at least adjacent buildings is recommended, but due to extensive and expensive workload, it is more or less often disregarded. Due to the lack of data on neighboring buildings, modeling of only the observed building is often resorted to.

Most buildings in the Zone A consist of massive longitudinal and orthogonal walls, masonry ceiling vaults or wooden ceiling beams and wooden roofs [1]. In Lower Town, the buildings are larger by the plan than in Upper Town, so the static systems are different. In the Upper Town, family houses and villas predominate, and in the Lower Town, multi-apartment buildings with mostly 7 to 9 residential units are the most common buildings.

The city of Zagreb has a moderate seismic hazard (expected peak ground acceleration in the return period of 475 years is ag = 0.25–0.28), but it is highly exposed. Preparedness and awareness for earthquakes in Zagreb (and in Croatia) are quite limited due to the low frequency of earthquakes. The last devastating earthquake hit Zagreb 130 years ago.

Many hospitals, schools, business premises, residential and government buildings, cultural institutions, monuments, churches and chapels are located in Zone A and are protected either as part of a historic urban complex or as individual heritage buildings per se. A total of 72% [1] of buildings in Zone A suffered major damage due to the earthquake; to compare the damage suffered by this area is almost proportional to the value of its cultural heritage. Zone B consists of a variety of urban patterns and a large number of immensely valuable buildings [12].

In addition to residential buildings under heritage protection, according to the World Bank report, in the cultural sector, the total number of affected buildings comprises 192 cultural institutions, 13 state heritage buildings, and 159 religious buildings across the City of Zagreb and the two counties. Since most of the buildings in this sector are very old, moderate to severe structural damage was sustained by 118 facilities (32%), and heavy structural damage was reported in 41 buildings (11%). Total damage to buildings and other physical assets is estimated at 1.38 billion EUR. The earthquake has diminished certain characteristic features of Zagreb's urban ambience. Damage has been done to unique stylistic and decorative elements of many of its historic buildings, and it has been necessary to remove many unstable features. Much of the damage involved the collapse of chimneys, gables and pediments, attics, prominent cornices and other elements.

3 The Post-earthquake Assessment of Building Stock in Zagreb

The first step in a complete post-earthquake buildings assessment was a rapid, preliminary assessment of the usability [13, 14] of all buildings damaged in the earthquake. In cases where it's needed, detailed assessment and available Non-Destructive Testing (NDT) assessment methods [15] should be applied. In Croatia, the rapid post-earthquake assessment consisted of a quick visual inspection of individual elements of the load-bearing structure, stating the appropriate degree of damage and deciding on the classification of the building into one of six possible categories: U1 Usable without limitations (Green label), U2 Usable with recommendations (Green label), PN1 Temporary unusable - detailed inspection needed (Yellow label), PN2 Temporary unusable – emergency interventions needed (Yellow label), N1 unusable due to external impacts (Red label) and N2 unusable due to damage (Red label).

The detailed assessment of the damaged buildings is regulated by the new public law [16], according to the procedures in the valid norms, Eurocode norms, and especially EN 1998–3 [17]. According to the new law "Law on the Reconstruction of earthquake-damaged Buildings in the City of Zagreb, Krapina-Zagorje County and Zagreb County (NN 102/2020)" [16] ultimate limit state return period can be different depending on the level of strengthening for old masonry buildings damaged in the recent earthquakes. Usually, the limit state of significant damage with a return period of 475 and limit state of damage limitation with a return period of 95 years has to be checked according to [17]. In the new law [16], the return period of 225 years which corresponds to a probability of exceedance of 20% in 50 years is introduced for a limit state of significant damage.

The usual detailed assessment consists of a concise visual inspection of all of the structural elements and details, an inspection of structural and non-structural damage, studying archive documentation, on-site and laboratory investigation of material properties, and writing a detailed report on the mentioned assessment procedures and results. All damage, structural and non-structural, is photographed and described in a final report. The MQI method which is explained in more detail in [18] and further developed in [19, 20], can be used when full NDT measurements should be applied. Based on the visual classification of the quality and regularity of the wall elements and mortar joints approximate relatively narrow range of values of material properties are given (shear strength, compressive strength and modulus of elasticity). Ultrasonic pulse velocity tester is also one of the NDT methods used for testing masonry. Based on the speed of the sound wave propagation through material modulus of elasticity can be calculated. Ultrasound measurements can be used for relatively new and compact masonry elements. Still, from the authors' experience, the results for material properties of the masonry walls are often unusable due to the cracked state of the element. The flat-jack method is classified as semi destructive as thin horizontal slots must be cut in masonry to insert thin steel (or aluminum) sheets. By applying a hydraulic pressure on the masonry through flat-jacks and by measuring deformations with linear variable differential transformers, insightful material parameters of masonry like compressive stress state and modulus of elasticity can be obtained. In combination with small horizontal jack, the initial shear strength of masonry without the influence of vertical stress can be found as well as the coefficient of friction. Another useful NDT method is experimental modal analysis. It uses natural low-frequency ambiental excitations which are collected through highly sensitive piezoelectric accelerometers. Collected data is then processed and final results are presented in the form of mode shapes of vibrations of a building with its natural periods and accompanying percentage of relative damping for each mode shape.

Eurocode 8 offers various linear and non-linear seismic analyses. However, for existing structures, non-linear static (pushover) analysis is recommended. When talking about existing structures, knowledge level about their details, material properties and geometry is important as it defines allowable seismic analysis and appropriate values of confidence factors. There are three levels of knowledge: KL1 (limited knowledge), KL2 (normal knowledge) and KL3 (full knowledge). Every knowledge level has its associated confidence factors, which are 1.35 for KL1, 1.2 for KL2 and 1.0 for KL3. Confidence factors are used to determine the design values of the material properties of an existing building. In other words, mean values of material properties obtained from in-situ tests are divided with an appropriate confidence factor. Such reduced material properties are then used in the calculation of the capacity of the building. Further, safety verifications are conducted where capacity obtained through pushover analysis is compared with appropriate seismic demand. Usually, geotechnical tests are not performed but based on empirical data, a category B foundation soil (deposits of very compacted sand, gravel or hard clay, at least several tens of meters deep) or category C (deep deposits of compacted or medium-compacted sand, gravel or hard clay with a thickness of several tens of meters to several hundred meters) can be assumed for the most parts of Zagreb.

In addition to the "traditional" inspection, case studies should be inspected from the air by an unmanned aerial vehicle (UAV) because that way damage can be observed to the main load-bearing structure and the building's roof structure. Decorative crosses, statues and reliefs must also be inspected. For digital preservation, the 3D model of buildings can be made based on photogrammetric images. As an addition to UAV, laser scanning can be extremely useful for thorough interior inspection, especially when the building in question has a complex geometry and parts hard to reach. With a compact imaging laser, load-bearing structure geometry and damage can be detected with high accuracy.

Depending on the possibilities in the future, a static analysis of the existing condition of the building should be made. In Croatia, the most often used software are 3Muri, Scia Engineer and Tower. It should be said that very often, unfortunately, non-linear static analyses are not used. After the performed analyses, the real damage can be compared with the damage distribution previewed in software. Description of building's capacity through displacement rather than forces allows us to understand better and accurately predict buildings response in the form of damage initiation and propagation throughout all phases of analysis until the formation of failure mechanism and collapse. After the full assessment, the usability of the building is evaluated and procedures for the renovation are given.

4 The Preservation of the Heritage Buildings with UAVs and Laser Scanners

The subjects of this paper are buildings in the city centre of Zagreb which experienced damage in the recent earthquakes and are under heritage protection. Every case study is very shortly presented with a brief description, photo and/or 3D model obtained by UAV photogrammetry or laser scanning.

The post-earthquake assessment was briefly explained and digitalization of heritage properties is shown on four different case studies. The focus is set on unmanned aerial vehicles (UAVs) and supporting equipment, as well as photogrammetry and laser scanning. The advantages and disadvantages of the new technologies which should complement traditional methods are given by [4].

Laser scanning was done in order to create a 3D point cloud of the building's interior and exterior. The laser scanner used in the case studies given below is a Leica BLK360 device, which is a compact imaging laser scanner that uses a 360° laser distance meter and high definition panoramic imaging. Collected data is then processed in the Cyclone Register 360 software package. 3D point clouds are delivered with millimeter accuracy with the help of three spherical, panoramic HDR cameras with a thermal imaging camera. The point cloud can further be used to make an accurate 3D model, to create a precise 2D floorplan or to convert it into a mesh for visualization. Every case study was filmed from the air and a digital copy of the outer façade was made.

4.1 Residential Building in the Lower Town (Klaićeva Street)

The building (Fig. 3) has a residential purpose and it is located in an area that is under cultural protection. It is made of unreinforced masonry with wall thicknesses from 60 to 45 cm. Floor structures and roofs are made of timber. In rapid assessment, it was classified as temporarily unusable due to the damage suffered.

Fig. 3. Drone photo and a point cloud of the building in Klaićeva

4.2 Residential Building in the Lower Town (Vinogradska Street)

This adjacent building (Fig. 4) is located in Zone B of Zagreb's historic urban complex. Originally the two-story building was built in 1896. and upgraded to a three-story building in 1938. reconstruction.

Fig. 4. Drone photo and section of a point cloud of the building in Vinogradska street

It consists of the basement, ground floor, first floor and attic. It is a residential building with a rectangular ground plan in the north-ssouth direction. The building's external dimensions are 14,3 × 12 m with massive longitudinal walls, masonry ceiling vaults in the basement, wooden ceiling beams and a wooden roof.

4.3 Residential Building in the Upper Town (Srebrnjak Street)

This case study (Fig. 5) is located in Upper Town, a part of Zone A of Zagreb's historic urban complex. It is a residential building with a rectangular ground plan attached to the symmetrically identical building. It consists of the basement, ground floor, first floor and attic. The building was built in 1932. while its original plan was never completed and part of the building was never built. It is a square-shaped building with a masonry construction with timber floors and a timber roof. The building was regularly maintained and suffered minor local damages in the latest earthquake.

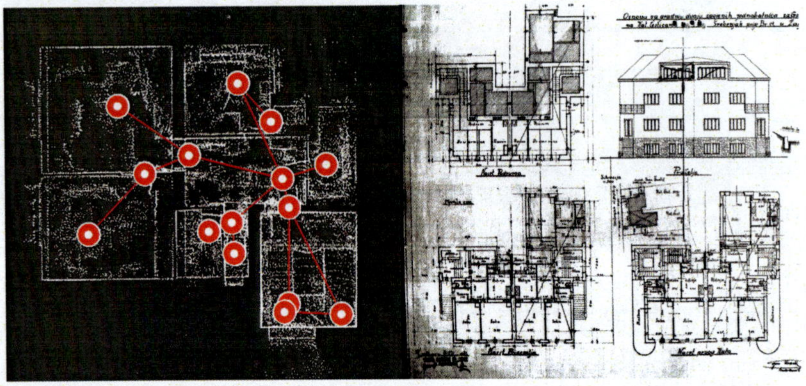

Fig. 5. Point cloud with scanning locations for one floor and archive ground plan of the building in Srebrnjak street

4.4 Public Building in the Lower Town (The Republic of Austria Street)

The next case study is a public building that is a part of the historical complex of the infantry barracks of Prince Rudolf and is located at Republic of Austria 18 street in Zagreb. The entire complex of Rudolf's barracks is protected as an immovable individual cultural property and is entered in the Register of Cultural Heritage of the Republic of Croatia. The building (Fig. 6) has a rectangular floor plan with dimensions of 25.18 m × 11.42 m and a height of approximately 15.50 m. The floor area of one floor is around 290.00 m^2, and the total gross area is around 1450 m^2. After the Zagreb earthquake, it was classified as temporarily unusable.

Fig. 6. Drone photo and a point cloud of the building in The Republic of Austria street

The laser scanning resulted in a point cloud with a precision of 3 mm and the facade of the building with external geometric contours was preserved for the future. In addition, the roof was filmed with a DJI Mavic 2 Pro drone and the point cloud from photogrammetry was done in the Agisoft Metashape software package.

5 Discussion and Conclusions

This paper briefly presents the post-earthquake damage assessment in the historic parts of Zagreb after the recent earthquakes on March 22, 2020. Although of a moderate magnitude, earthquakes caused much damage, significant financial loss and revealed many flaws in the built heritage that residents, decision-makers, and the professional and scientific community will have to deal with for many years to come. The earthquake extensively damaged older masonry structures built before any enforced seismic codes. Therefore, much damage was observed in the historic part of the city.

Almost the whole city center is under some level of cultural heritage protection. The protected historical urban complex is, therefore, the part of Zagreb that is the most recognizable and gives the city an identity. The general conclusions regarding the analyzed part of the city are that the damaged buildings are generally old and built before the seismic codes.

The paper analyzes the preliminary data of the usability of the buildings in the heritage protected areas by field engineers and is supplemented with the four case studies where, in addition to regular post-earthquake assessment, point clouds and digital twins of the protected buildings are shown. The basic idea was to show the procedures and possibilities for the preservation of cultural heritage buildings with state-of-the-art technology. The case study buildings should represent a good starting point for further analysis and as a digital twin for different types of users in the renovation process.

As Croatia was hit by two severe earthquakes last year, the knowledge of build back better is fully appreciated. That means that sustainable materials and innovative concepts [21–24] should be used and energy efficiency ensured [25, 26]. Preservation through new technologies should be included in the whole process. A compilation of all the assessment data can and should be obtained to get a unique view of the seismic vulnerability of existing structures and not only to the architectural preservation of the built heritage.

Acknowledgements. The Croatian Science Foundation funded this research through the ARES project: Assessment and rehabilitation of existing structures – developing of contemporary methods for masonry and timber structures (UIP-2019-04-3749).

References

1. Stepinac, M., et al.: Damage classification of residential buildings in historical downtown after the ML5.5 earthquake in Zagreb, Croatia in 2020. Int. J. Disaster Risk Reduction **56**, 102140 (2021)
2. ARES Project official website. https://www.grad.hr/ares/
3. Atalić, J., Novak, M.Š, Uroš, M.: Seismic risk for Croatia: overview of research activities and present assessments with guidelines for the future. Gradjevinar **71**(10), 923–947 (2019)

4. Stepinac, M., Gašparović, M.: A review of emerging technologies for an assessment of safety and seismic vulnerability and damage detection of existing masonry structures. Appl. Sci. **10**, 15 (2020)
5. Ortega, J., Vasconcelos, G., Rodrigues, H., Correia, M., Ferreira, T.M., Vicente, R.: Use of post-earthquake damage data to calibrate, validate and compare two seismic vulnerability assessment methods for vernacular architecture. Int. J. Disaster Risk Reduction **39**, 101242 (2019)
6. Sánchez Rodríguez, A., Riveiro Rodríguez, B., Soilán Rodríguez, M., Arias Sánchez, P.: 9 - Laser scanning and its applications to damage detection and monitoring in masonry structures. In: Ghiassi, B., P.B.B. T.-L.P., Lourenço, D.M.S. (eds.) Woodhead Publishing Series in Civil And Structural Engineering, Woodhead Publishing, pp. 265–285 (2019)
7. Manyika, J., Chui, M., Miremadi, M., Bughin, J., George, K., Wilmott, P., Dewhurst, M.: A Future That Works: Automation, Employment, and Productivity. https://www.semanticscholar.org/paper/A-future-that-works%3A-automation%2C-employment%2C-and-Manyika-Chui/e3ac9558f18234cdf92b730a7386_16d446a1af
8. McKinsey Global Institute. Global Media Report 2014. https://www.mckinsey.com/ndustries/technology-media-and-telecommunications/our-insights/global-media-report-2014
9. Lulić, L., Ožić, K., Kišiček, T., Hafner, I., Stepinac, M.: Post-earthquake damage assessment—case study of the educational building after the Zagreb earthquake. Sustainability **13**, 6353 (2021)
10. Fagundes, C., Bento, R., Cattari, S.: On the seismic response of buildings in aggregate: analysis of a typical masonry building from Azores. Structures **10**, 184–196 (2017)
11. Senaldi, I., Magenes, G., Penna, A.: Numerical investigations on the seismic response of masonry building aggregates. Adv. Mater. Res. **133–134**, 715–720 (2010)
12. Republic of Croatia Ministry of Construction and Physical Planning, Proposal of the Long-Term Strategy for Mobilising Investment in the Renovation of the National Building Stock of the Republic of Croatia, Zagreb (2014)
13. Uroš, M., et al.: Post-earthquake damage assessment of buildings – procedure for conducting building inspections. Gradjevinar **72**(12), 1089–1115 (2020)
14. Stepinac, M., Rajčić, V., Barbalić, J.: Inspection and condition assessment of existing timber structures. Gradjevinar **69**(9), 861–873 (2017)
15. Stepinac, M., Kisicek, T., Renić, T., Hafner, I., Bedon, C.: Methods for the assessment of critical properties in existing masonry structures under seismic loads-the ARES project. Appl. Sci. (Switzerland) **10**, 5 (2020)
16. Law on the Reconstruction of Earthquake-Damaged Buildings in the City of Zagreb, Krapina-Zagorje County and Zagreb County (NN 102/2020)
17. Eurocode 8: Design of structures for earthquake resistance - Part 3: Assessment and retrofitting of buildings (2005)
18. Borri, A., Corradi, M., Castori, G., De Maria, A.: A method for the analysis and classification of historic masonry. Bull. Earthq. Eng. **13**(9), 2647–2665 (2015)
19. Borri, A., Corradi, M., De Maria, A., Sisti, R.: Calibration of a visual method for the analysis of the mechanical properties of historic masonry. Procedia Struct. Integrity **11**, 418–427 (2018)
20. Borri, A., Corradi, M., De Maria, A.: The failure of masonry walls by disaggregation and the masonry quality index. Heritage **3**, 4 (2020)
21. Fortunato, G., Funari, M.F., Lonetti, P.: Survey and seismic vulnerability assessment of the Baptistery of San Giovanni in Tumba (Italy). J. Cult. Herit. **26**, 64–78 (2017)
22. Stepinac, M., Šušteršič, I., Gavrić, I., Rajčić, V.: Seismic design of timber buildings: highlighted challenges and future trends. Appl. Sci. **10**, 4 (2020)
23. Funari, M.F., Mehrotra, A., Lourenço, P.B.: A tool for the rapid seismic assessment of historic masonry structures based on limit analysis optimisation and rocking dynamics. Appl. Sci. **11**, 3 (2021)

24. Funari, M.F., Spadea, S., Lonetti, P., Fabbrocino, F., Luciano, R.: Visual programming for structural assessment of out-of-plane mechanisms in historic masonry structures. J. Build. Eng. **31**, 101425 (2020)
25. Milovanović, B., Bagarić, M.: How to achieve nearly zero-energy buildings standard. Gradjevinar **72**(8), 703–720 (2020)
26. Valluzzi, M.R., et al.: Nested buildings: an innovative strategy for the integrated seismic and energy retrofit of existing masonry buildings with CLT panels. Sustainability **13**, 3 (2021)

The Bell – Tower of Saint Barbaros in the Medieval '*Borgo*' of *Potamos* in Corfu: Historical Documentation and Multi-disciplinary Proposal for its Preservation

Evi Polymeri[1](✉), Charalampos Mouzakis[2], Athanasios Efstathiou[3], and George Pantazis[4]

[1] Department of Preservation of Antiquities and Works of Art, Faculty of Environment, Ionian University, Zakynthos, Greece
epolymeri@ionio.gr

[2] School of Civil Engineering, National Technical University of Athens, Athens, Greece

[3] Dean of the Faculty of Humanities, Ionian University, Corfu, Greece

[4] School of Rural and Surveying Engineering, National Technical University of Athens, Athens, Greece

Abstract. Having been built at the beginning of the 17th century the historic bell tower of Saint Barbaros in Corfu belongs to the post-byzantine Church of The Mother of God of the Life-Giving Spring and St. Barbaros. It is considered to be a representative case of a surviving monument with occidental, Venetian architecture elements, techniques and masonry in the Ionian Islands. There are severe construction and static issues identified and today's state of the monument is considered to be at risk. A study customizing all the components through non-destructive techniques, providing a preventive maintenance and assessment proposal for restoration and interventions is our first priority to save the monument. The bell tower belongs to the category of tower constructions. Through preliminary 'in situ' examination and the analysis of the documentation we come to the conclusion that the constructions' masonry belongs to different historical phases and there is severe damage caused mainly by natural disasters such as earthquakes, but at the same time by human activity such as periodic interventions. An attempt is made to decrypt the direct and indirect historical documentation for the identification of evidence regarding the hidden construction phases and severe damage that the building suffered. Historically the destructive factors mentioned induced its partial collapse in different historical periods. The answer to the vulnerability of this cultural heritage structure is a sustainable preservation of St. Barbaros bell tower in Corfu through a Transdisciplinary Project and Cooperation for effective Protection of Cultural Heritage in the Ionian Islands. Having considered the importance of the new findings and evidence according St. Barbaros the Myrrh-Effusing, connected to the cult of the homonymous saint and the post-byzantine Church of The Mother of God of the Life-Giving Spring and St. Barbaros, we ring the alarm bell for efficient preservation of the bell tower, a monument of cultural heritage of the Ionian Islands.

A. Osman and A. Moropoulou (Eds.): *Advanced Nondestructive and Structural Techniques for Diagnosis, Redesign and Health Monitoring for the Preservation of Cultural Heritage*, SPM 16, pp. 178–193, 2022.
https://doi.org/10.1007/978-3-031-03795-5_7

Keywords: Bell-tower · St. Barbaros · Structural system · Masonry · Cultural heritage · Preservation · Transdisciplinary project · Corfu

1 Introduction

For centuries, bell towers played a crucial role in the religious life of people and at the same time served defense strategies to protect Corfu island against the Ottoman sieges at the end of 16th–18th cent. Having been used from ancient times as *semantro* [3], the high constructions were conspicuous in the distance between the other buildings serving at the same time as sailors' orientation. While assisting the community's needs the bell towers were part of people's everyday life, of their joys and their griefs, until today. Considered as the highest of the kind in Corfu, this particular bell tower reaches 35 m high and it might be spotted with the naked eye in the wider area of Potamos suburb, the Town, the New Fortress and the villages around.

The bell tower of St. Barbaros (Fig. 1, 2.) of The Mother of God of the Life-Giving Spring Church is located in 'borgo' Potamos, a large suburb of Corfu Town situated 4 km north. There is a river crossing through the area, the largest and most important in Corfu. The two banks of the river are joined by a stone bridge dated to the first half of the 19th cent. Potamos comprises mainly rural areas and extends to the sea in Alykes.

Fig. 1. Bell tower of St. Barbaros, church of the mother of god of the life-giving spring. (Phot. Kastamonitis, E.).

Fig. 2. Bell tower of St. Barbaros, 1839–40, Gennadius library - American school of classical studies at Athens (ASCSA).

The tower constructions in Corfu mostly dated from the post-Byzantine period. It is noteworthy to mention that historic bell towers are located in Corfu scattered through the island, with different density and characteristics. In fact, we have not found yet architectural plans or other information about the way they were built or about their dating. Although, there are differences in their morphological and typological characteristics. These constructions mostly are considered to be 'independent' and admittedly their main feature, the height, offers to the construction a unique peculiarity.

The bell tower of St. Barbaros was named by the homonymous *martyr* whose relics were carried by a ship which passed from the mouth of the river to get water supplies. There the local people came as pilgrims and they sought to be healed. The Christian Orthodox cult of St. Barbaros was cultivated considering that Barbaros came from Africa or the Arabian world. According to D. Zakythinos' [13] and H. Delehaye's [12] studies, there are four different versions of the Martyrdom of St. Barbaros (Logos of Constantine Acropolites [2], BHG 220 [18], the Bulgar version [12] etc.) and different historical periods. Having considered Saints' Synaxarion [6] we are informed that in 1571, while travelling by ship on their way to Venice a Venetian military man named Sklavounos, who took part in the naval battle of Naupactus, suddenly fell ill with a deadly disease. Having had a vision of St. Barbaros, who invited the patient to worship the martyr's tomb, with the aim of being healed, the sick soldier reached the Saint's tomb, worshiped reverently and immediately was healed. Willingly to honor St. Barbarus the Myrrh-Effusing, he recovered his relics, in order to transport them to Venice. On the 15th of May, while sailing, in need for water supply, the ship stopped near to Corfu town, passing through Potamos River. There, a paralyzed young man was cured. After the miraculous event the local people decided to add a new name to their church after St. Barbaros the Myrrh-Effusing and celebrate on the 15th and 23^{d} of May. In the festive icon of St. Barbaros in the Southern part of the church we consider, there are presented the first known artistic depiction of the village Potamos and saints' miracles signed by the artist Georgios Chrysoloras, dated on 1720 (Fig. 3). The bell tower of the church Mother of God of the Life-Giving Spring and St. Barbaros the Myrrh-Effusing, probably built in 1605, is rendered in an almost faithful form (Fig. 4).

Having analyzed another important source, such as Vincenzo Coronelli's book published in 1699, we found 'Capo Barbaros' (Cape Barbaros) location mentioned in the text. In the following pages there is a map of Corfu island (Fig. 5) testifying both 'St. Barbaros' and 'Capo Barbaros'(Fig. 6). We believe that the location might be identified with today's area between 'Avlaki' and 'Eremitis', in front of 'Caparelli' (islet with the lighthouse). While considering Coronelli's map we might come to the conclusion that St. Barbaros was venerated not only in Potamos, but generally in Corfu island and the places where the ship anchored were named by the Saint (the same as with Saint Nicolas in Corfu island).

Fig. 3. The festive icon of St. Barbarus the myrrh-effusing, tempera painting, Chrysoloras, G. 1720. (Phot. Polymeri, E.).

Fig. 4. Detail of the festive icon of St. Barbarus the myrrh-effusing, tempera painting, Chrysoloras, G., 1720. (Phot. Polymeri, E.).

Fig. 5. Isolario dell' p. Coronelli, 1696, pp. 68–69

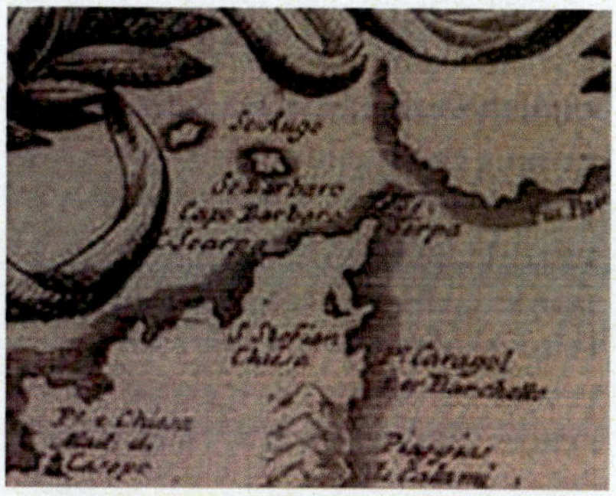

Fig. 6. Detail, Isolario dell' p. Coronelli, 1696, p. 69.

The location 'St. Barbaro' is testified too, by Charles Louis Fleury Panckoucke's (1780–1844) map of Corfu island and Territory of Butrinto (Buthrotum), Siege 1798–99, published in1818 (Fig. 7).

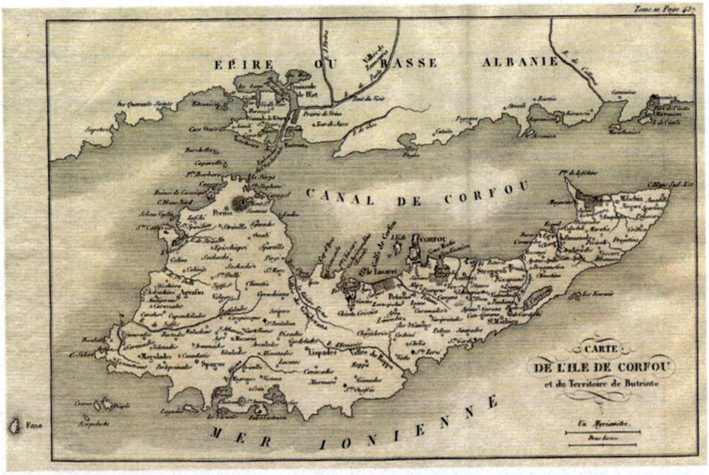

Fig. 7. Map of Corfu Island and Territory of Butrinto (Buthrotum), Siege 1798–99, Charles Louis Fleury Panckoucke, 1818.

2 Methods and Materials

The bell tower of St. Barbaros Fig. 1, 10 in Potamos suburb was built from limestone produced by local quarries on a mighty square pyramidal base with a floor plan of approximately 5 m × 5 m. The total height to be reconstructed from the findings was thus once approximately 35 m. Having been identified there are different types of assessments in the construction – for example in some part lime stone, in other there are bricks and plaster, lime mortar and at the base, the oldest, clay mortar etc. According to the sources the mortar was prepared adding egg yolk to the mixture. In the middle of the 19th century we find that the coating of the exterior facade of the bell tower was painted in *celeste blue* dye. The fact is documented by the paintings of Edward Lear (1855) (Fig. 8) and Angelos Giallinas (1900) (Fig. 9). The paintings of both artists are considered by art critics as mostly realistic and naturalistic.

A visually accentuating, striking profile cornice is executed above the pyramid step of the bell tower. Particularly, noteworthy in its west facade, in local yellowish-white coarse limestone, designed baroque passage portal in the style of the neo-Renaissance, the one indicated here on the keystone. There is an engraved inscription up to the entrance, above the pyramidal step cornice of the bell tower: 'Built in 1605' (Fig. 11). The date was placed by the priest of the parish in the 1960s. The key of the cornice in the entrance gate has an engraved Latin cross and immediately after embossed decoration. For the construction of the staircase in the interior of the bell tower were used mainly wood and some metallic parts.

Fig. 8. St. Barbaros bell tower painted in '*celeste blue*' dye, Edward Lear, 31.12.1855. Private Collection.

Fig. 9. St. Barbaros bell tower traces of the '*celeste blue*' dye, Angelos Giallinas, 1900. Private Collection.

Fig. 10. Bell tower of Saint Barbaros, church of the mother of god of the life-giving spring and St. Barbaros (Phot. Kastamonitis, E.).

Fig. 11. Entrance of St. Barbaros bell tower, Potamos, Corfu. (Phot. Polymeri E.).

According to the date of construction, through our research, we come to the conclusion that the bell tower didn't exist before 1571[1]. There is a map by an unknown author dated back at the beginning of 16th cent. indicating that during the Ottomans' siege in 1537 (Fig. 12), the church of St. Barbaros was displayed without the bell tower, which means that the high construction was built later, strengthening our theory.

Fig. 12. Unknown, 1537, Corfu in the period of the Ottoman Siege, 1537, in circle 'borgo' Potamos.

At the end of the 16th– early 17th century in order to strengthen its presence and claim its supremacy the Most Serene Republic of Venice proceeded in the construction of great architectural works such as palaces, buildings and very high bell towers in its capital, Venice, for example San Marco's bell tower, and at the same time in other cities of Italy, as well as the Ionian Islands, such as Corfu, Zakynthos and Cephalonia. To strengthen the same purpose the bells were carefully selected, so the result would be auspicious and in case of enemy threat to signal alarm. An example is considered to be the case of the bell tower of St. George of the Greeks Church in Venice (Fig. 13) which was erected in September 1587, a few years before that of St. Spyridon in Corfu. Having had regard to the bold work of the Venetians, Marmoras A. notes: Corfu may compete worthily the most beautiful bell towers of Italy [25]. Having considered the architectural category of bell tower constructions we might argue that it is probably the oldest of its kind appearing in Corfu at the end of 16th century with reference to the bell tower of Saint Spyridon's Church in the Old Town of Corfu (year of construction 1590) [1] (Fig. 14).

The symmetrical shape of St. Barbaros bell tower (Fig. 15) in and its strong vertical axis lead to almost "free building" with architectural and static autonomy. This means that no fixed rule is observed in terms of its location and the way it is arranged seems to follow the possibilities offered by the surrounding area of the church serving its functional needs. It is located to the northwest of the church in the center of Potamos' suburb, it is a very high and closed building. This 'opus maximus' project construction requires advanced knowledge and know-how to be designed and executed at the same

Fig. 13. Bell tower of St. George Greek church, Venice, Italy.

Fig. 14. St. Spyridon's bell tower, 1857. Phot. Major John Davenport Shakespear, RHA.

Fig. 15. View of St. Barbaros' bell tower. Phot. Alois Beer, 1886–1890.

time. It also has vents, balconies and a '*cupola*' end. On the north side the foundation rests on a retaining wall, and it is not clear if it was built together with the bell tower. However, this wall is made of stone. In the bell towers of the Ionian Islands, the ratio 1: 6 to 1: 8 (ratio of base side to total height) is applied, while in mainland Greece the ratio decreases respectively to 1: 4.

Apart from the trunk, which is made of stone and testifies to the date of its construction, the rest of the bell tower is made of brick and we find the existence of at least two subsequent interventions. Due to the development of the phenomenon of corrosion, the mortar is fragile and part of the binder material has been detached together with part of the bricks, creating a gap in the masonry, thus affecting the statics of the building.

The tower system belfry of St. Barbaros is divided into three parts, the lower part which is the base in the shape of a truncated pyramid, followed by the trunk, which includes, inside it, the winding staircase and the upper part with the insert of the first balcony in which the bells with the capitals and a second balcony are placed on top, perforated on all four sides. The third and last part is the crown of the bell tower, it has the shape of a 'cupola' with the insertion of a drum and a pillar. An iron cross with a weather vane is placed at the top of the construction[2]. That part is dated to the modern period and we consider that in different historical periods damage was caused and it collapsed, (we are not aware how many times) mainly due to natural disasters such as earthquakes. Seismic parameters and remarks according the most destructive earthquakes in Corfu and the Ionian Islands area recorded mainly by J. Partsch (Table 1) in his book 'The island of Corfu: geographical monography' [31].

According to the surviving base of the bell tower Fig. 11, we strongly believe that it belongs to the first phase of the monument, which might be the beginning of the 17th

Table 1. Seismic parameters and remarks according the most destructive earthquakes in Corfu and Ionian Islands area by J. Partsch [31].

Year	Month/date	Magnitude	REMARKS
1650	26th February	6.0	Destructions/St. Athanasios church at the homonymous bastion, Town and villages.
1666	November	6.2	Extensive earthquake.
1674	16th January	6.5	Extensive destructions.
1704	11th November	6.6	
1732	March	6.5	Very Strong. Destructions/walls, parts of the fortifications.
1743	20th February	7.1	Very Strong. Destructions/buildings: Latin Archbishopric Residence, the Palace of '*Proveditor General da Mar,* and bell tower of 'Annunziata'.
1745	3d April	6.2	Strong. Destructions/govern. buildings
1756	13th February	6.3	Building destructions
1767	22nd July	7.2	Very strong. Major disasters/buildings, town and villages around
1773	May	6.4	Very Strong. Destructions/buildings
1786	End of January-5th of February	6.6	Very strong, 120 dead, and buildings collapsed, such as the Palace of the governor.
1809	3-4th May		
1813	December		

century. The large and smaller stones as well as the style of the masonry and the mortars refer to constructions of the same period (see fortification works *'fronte bastionato'* planed by the great Venetian architect Michele Sanmicheli [1, 33]). Today's coatings and interventions with modern mortars are due to the occasional repair work without any particular plan. The part built by bricks seams to accord to the modern period.

3 Church St. Barbaros and of the Mother of God of the Life-Giving Spring

The Orthodox church of St. Barbaros and of The Mother of God of the Life-Giving Spring, in the form that survives today, is a simple single-aisle basilica plan, timber-roofed church of purely so-called Ionian type, formed under the influence of the western church. Its plan (Fig. 16.) consists of an elongated rectangle, slightly crooked, with a protruding semicircular arch of the sanctuary. The total length of the Northeastern side of the church is approximately 21.64 m., 22.08 m., the Southwestern side 22 m. and its width reaches in the West, 9.36 m. and in the East, on the side of the sanctuary approximately 12.15 m. The proportions of the sides of the rectangle are 1:2, characteristic of the single-aisle churches of Corfu during the Venetian period.

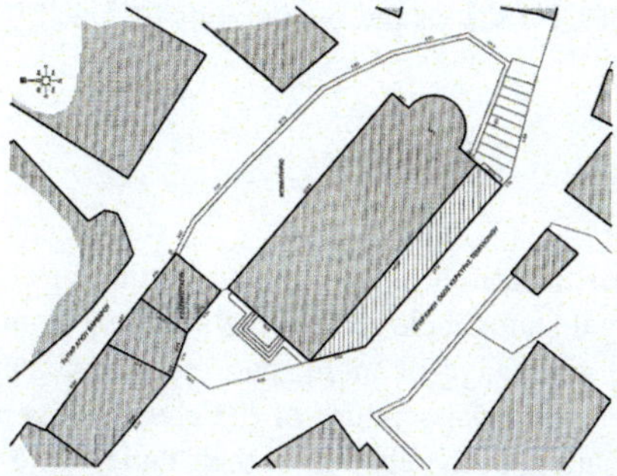

Fig. 16. Plan of the church and the bell tower (1:200)

Fig. 17. Detail of the old wall at the foot of the bell tower. (Phot. Polymeri, E.).

During the reconstruction works inside the church which took place in 1999–2000, by chance fragments of frescoes covering part of the hereon on the east wall of the sanctuary were revealed. The mural depicts scenes in which the so-called liturgical cycle unfolds, including, among others, the representation of the Communion of the Apostles and the Liturgy of the angels. At the top depicts in medal Christ Enthroned in Glory and below the Holy Shroud with Jesus Christ up to the base of the neck and rich hair. This depiction is considered rare and mainly used in conservative provinces' iconographic program. The monumental painting of the church reminds us of an earlier phase of the church, in fact it proves that it was larger in size and that there was financial prosperity so that outstanding artists could be assigned the work of designing and finishing the painting of the decoration. The technique of the fresco fragments is of fine quality and the colors are quite vivid. But there are serious problems related to the high levels of dampness. The frescoes revealed probably belong the earlier phase of the monument. Today parts of the mural are at risk.

On the other hand, references to archival documents of the 15th century of the church of The Mother of God of the Life-Giving Spring and St. Barbaros reinforce the view that the church not only pre-existed, but was also quite richly imposing in the space.

The church, as an autonomous building, together with the same autonomous bell tower in the heart of the 'borgo' Potamos, comprise a high quality and very interesting architectural ensemble. In the courtyard of the church, the floor is made of stone, and has large limestone slabs of local origin. On the surface of the stone slabs we observe, by country, engraved inscriptions or monograms, and in the center, we locate sloping metal handles. During the research, it was found that, in an earlier phase, the stone slabs covered the graves that had been placed in the courtyard of the temple on the Northern part of the courtyard of the temple. Later, mainly for reasons of hygiene, burial was forbidden in the areas around the church.

Externally, to the NE, the church and its bell tower are surrounded by a high wall which follows the slope of the ground and rises accordingly (Fig. 17). After an 'in situ' investigation, it was found that part of the preserved wall reflects earlier phases of the monument, in fact, it seems that there were many. The diversity and variety of building materials and mortars found in the masonry testify to some phases of the monument. The church is recorded, according to archival sources, in the published list of churches of Corfu by Spyridon Papageorgiou 1920 [30], as the monuments included in this list are dated before 1750.

4 'Jus Confraternita'

The ownership state[3] of the church of St. Barbaros and of The Mother of God of the Life-Giving Spring in terms of its legal status, belonged to the religious Confraternity [30], which means it was under the patronage of more than one prominent family. The decisions on the financial, land and property issues of the church in general, the religious liturgical issues and other matters of the church. In their duties were included the construction and other works for the preservation of the church and the bell tower, the purchase and care of its equipment, as well as the care for its regular function. As for the property issues, we know from the notarial documents and manuscripts that the church was quite rich financially and in property. According to archival sources (Fig. 18, a, b.), the church owned areas with olive trees, land, buildings, houses, etc. which were handled by the respective administration of the patronage.

Having had the inheritance rights in possession the members of the Churches' Council could legally pass it to their descendants, or appoint successors without seeking the approval of the ecclesiastical authority. The Confraternities were built on the acquisition of patronage rights of a church, equal to all the patrons.

Notes: 1. *Recent finding confirms that the worship of the Saint was already known and flourishing in Potamos suburb, therefore the passage of the Saint from Corfu and the borgo of Potamos can be placed by the end of the 15th - beginning of the 16th century, earlier than 1571.* 2. *In modern times a metal ladder and a metal lightning conductor were added to prevent the destruction of the bell tower.* 3. *It should be noted that the Greek Church remained in Corfu under the spiritual leadership of the Great Protopapas (Μεγάλου Πρωτοπαπά), surrounded by episcopal jurisdiction. The ownership state of churches and monasteries in the Venetian-occupied areas of the Ionian Islands such as Corfu, Zakynthos, Cephalonia, Lefkada, starting from the first years of the Venetian conquest, were divided into four categories: a) di Confraternita, which means the ownership belonged to many families or even guilds (there were two categories of Confraternities: 1. patronage and 2. ktetoric* [17]*). b) Jus Patronato Privato, the church founders, the*

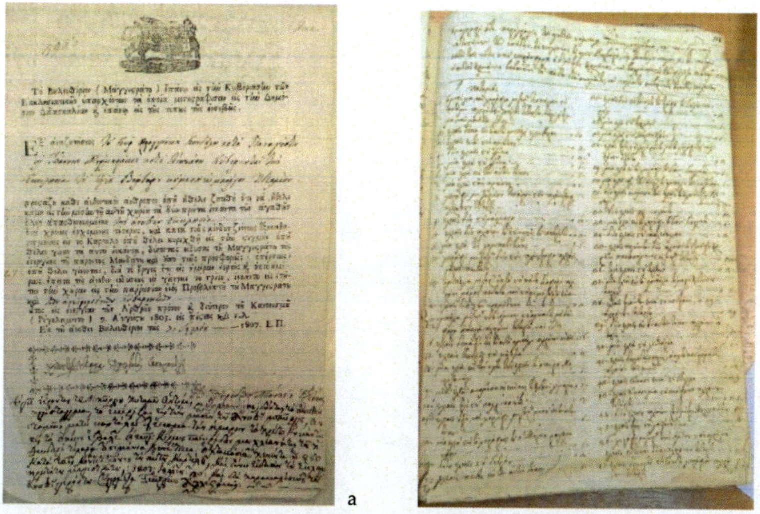

a b

Fig. 18. a), b). Church documents/Πρωτοπ., Φ/466., Δ.218, G.S.A (Γ.Α.Κ.)

ownership belonged to individuals, c) the monastic ones and d) Jus Patronato Pubblico the ownership belonged to founders of Public Law [19, 20]. *The detailed presentation of these distinctions is registered by the Proveditor General da Mar, Augustinos Sagredos, on August 26, 1754. In the list of churches described, we find the church of St. Barbaros and of The Mother of God of the Life-Giving Spring in 'borgo' Potamos.*

5 Assessment of the Seismic Response of the Bell Tower

The exact geometry of the bell tower, as determined through survey and photogrammetry, as taken into account for the investigation of the seismic response of the tower (Fig. 19). Two crucial elements affecting the response of the structure are the unequal height of its foundation, as well as its being in full contact with the neighbouring building (Fig. 20). Furthermore, it was observed that there are two distinct material phases above the upper face of the pyramid. It should be noted that the vertical projections at the corners of the facades were constructed at a later time, since the masonry is in front of the stone masonry of the upper part of the pyramid. The initial configuration of the bell tower did not include these projections, leading to the conclusion that the initial tower of 1605 collapsed during a strong earthquake, possibly due to the event of 1743 [35]. The damage occurred above the pyramid, and afterwards the tower was rebuilt in its present form with the solid bricks.

Seismic hazard analysis was carried out and time histories were selected for the numerical analysis of the bell tower at Ultimate Limit State (ULS) and Near Collapse (NC) limit state. A 3D model was developed (Fig. 3), the solid elements of which had dimensions equal to the respective dimensions of the masonry unit plus the mortar joint

Fig. 19. The bell tower today

Fig. 20. Meshed 3D model for numerical analysis

Fig. 21. Foundation level

thickness, considering the masonry as homogenous material. The developed 3D model is shown in Fig. 21. For ULS, the failure criterion was based on principal stresses, while for NC the criterion was the interstorey drift. An exceedance of masonry tensile strength was shown at various locations, with damages concentrated at the base and top of the arches of the bell tower.

Based on the above, an in-depth study of the monument is proposed.

6 Conclusions

The answer to the vulnerability of this cultural heritage structure is a sustainable preservation of St. Barbaros bell tower in Corfu through a Transdisciplinary Project and Cooperation for effective Protection of Cultural Heritage in the Ionian Islands. Having considered the importance of the new findings and evidence according St. Barbaros the Myrrh-Effusing, connected to the cult of the Christian martyr and the post-byzantine Church of The Mother of God of the Life-Giving Spring and St. Barbaros, *we ring the alarm bell for efficient preservation of the bell tower, a monument of cultural heritage of the Ionian Islands and an in-depth study of the monument is proposed.*

References

1. Agoropoulou-Birbili, A., The architecture of the Old Town of Corfu (NTUA), Athens, pp. 264–267 (1976)
2. Papadopoulos-Kerameus, A.: '*Konstantinou tou Akropolitou logos eis ton agion Barbaron*', Analects of Jerusalem Stachyology, vol.1, St. Peters-bourg, typograph. V. Kirsvaum, pp. 405–420 (1891)
3. Barla, C.N.: Morphology and developments of the Byzantine bell towers, Library of the Archaeological Society at Athens, No.45, Athens, pp. 46–49 (1959)
4. Binggeli, A., Métivier S.: Les nouveaux martyrs dans l'hagiographie byzantine, Proceedings of the 23rd International Congress of Byzantine Studies, Edit. Bojana Krsmanović, Ljubomir Milanović, Assist. Edit. Pavlović, B., Belgrade, pp. 35–36 (2016)
5. Bishop of Fanarion Agathangelos, Maltezou, Ch., Morini, E.: The Holy relics of saints of the Eastern, Saint Barbarus, the Myrrh-Effusing, Orthodox Church in Venice, Athens, pp.119–126 (2005)
6. Ακολουθία και βίος του οσίου πατρός ημών Βαρβάρου του Πενταπολίτου Ψαλλομένη τη ιε' Μαΐου Νυν το δεύτερον εκδιδομένη μετά της εικόνος του Αγίου Υπό Ιωάννου Α. Αλεξάκη, Εν Κερκύρα, Εκ του Τυπογραφείου "Ο Κοραής", Ι. Ναχαμούλη (1886)
7. Cicogna, A.E.: Delle inscrizioni veneziane raccolte ed Illustrate, vol.2, Venezia, MDCCCXX/[1820], santo Barbaro martire, p. 372
8. Concina, E.: "San Marco, la cittadella e la città", Corfu: History, Urban Space and Architecture, 14th-19th Cent., Corfu, pp.32–37, 151/fig.219 (1994)
9. Coronelli, V.: Descrittione geografico-historica, sacro-profana, antico-moderna, politica, naturale, e poetica, mari, golfi, seni, piagge, porti, barche, pesche, promontorj, monti, boschi, fiumi ... ed ogni più esatta notitia di tutte l'isole: coll'osservationi degli scogli sirti, scagni, e secche del globo terracqueo: aggiuntivi anche i ritratti de' dominatori di esse: ornato di trecento-dieci tavole geografiche, topografiche, corografiche, iconografiche, scenografiche ... a' maggiore dilucidatione, ed uso della navigatione, et in supplimento dei XIV volumi del Bleau: tomo II dell'Atlante veneto, Venice, pp. 167, map pp. 168–9 (1699)
10. Ćurčić, S.: Church towers, Annexes and Addition in Early Christian, Byzantine and Post-Byzantine Churches, Seminar 2 proceedings, Archaeological Museum Thesalonike, University Studio Press, pp. 103–111 (2008)
11. Da Costa-Louillet, G., Barbaros, V.S.: Saints de Grèce aux VIII e, IX e et X e siècles, Byzantion, vol. 31, no. 2, Peeters Publishers, pp. 309–313 (1961). http://www.jstor.org/stable/44201965
12. Delehaye, H.: Les actes de St.Barbarus, *Anal.Boll.*T. *XXIX*, Bruxelles, pp. 273–301 (1910)

13. Zakythinos, D., Barbaros, S.: In memory of K.Amarantos: 1874–1960, Honorary vol., Athens, pp. 438–453 (1960). Zakythinos, D., Αι ιστορικαί τύχαι της Επτανήσου και η διαμόρφωσις του Επτανησιακού πολιτισμού, Επιστημονική επετηρίς της Φιλοσοφικής Σχολής του Πανεπιστημίου Αθηνών, Περίοδος Β' ΤΟΜ. 15 (1964–1965), Athens, pp. 308–309
14. Zivas, D.: The bell towers (Part IV, chapter 2.), The Architecture of Zakynthos 16th–19th cent., Technical Chamber of Greece (TEE-TCG), pp.141–151 (2002)
15. General State Archives, Corfu/Γενικά Αρχεία του Κράτους, Αρχεία Κέρκυρας, Church documents, Πρωτοπ., Φ/466, Δ.218.
16. Gulielmo de Angelis d'Ossat (translation by Pandis N.S.), Efimeris Ionion, Sunday, XX, pp. 3, 28 June 1942
17. Horden, P.: The Confraternities of Byzantium. In: Horden, Hospitals and Healing from Antiquity to the Later Middle Ages, Book Part. II, Aldershot [u.a.], pp. 25–45 (2008)
18. Kazhdan, A., Talbot, A.M., Research Assist. Alexakis, A., Efthimiadis, S., McGrath, S., Sherry, L.F., Zielke, B.: 'Doumbarton Oaks Hagiography Database', Doumbarton Oaks Trustees for Harvard University, Washington, p. 7 (1998)
19. Karydis, S.C.: Collective patronage and the question of the origin of the Orthodox religious Confraternities in the Latin-Greek area, honorary vol., Studi di onore del Prof. Giorgio Fedalto, Hellenic Institute of Byzantine and Post-Byzantine Studies in Venice, Athens-Venice, pp. 237–258 (2016)
20. Karydis, S.C.: Orthodox confraternities and confraternity churches in Corfu (15th–19th c.): a contribution to the ecclesiastical history of the Ionian Islands, Athens (2004)
21. Karydis, S.C., Dimoulas, G., Poulis, S.: Places of worship and clergy in Corfu of the 17th century. Archival items, Corfu Reading Society, Corfu (2018)
22. Koumanoudis, I.: Corfiot bell towers. Zygos **11**, 36–52 (1966)
23. Kouris, S., Karaveziroglou-Weber, M.: Structural systems of masonry bell-towers and their earthquake response. Κουρής, Σ., Σεισμική Συμπεριφορά πυργωτών κατασκευών στην Κέρκυρα, Θεσσαλονίκη (20210)
24. Lampsidis, O.: Une nouvelle version de la vie de saint Barbaros, 'Πλάτων', Εταιρείας Ελλήνων Φιλολόγων, vol. IH, No. 35–36, pp. 40–56 (1966)
25. Marmoras, A.: Della historia di Corfu, descritta da Andrea Marmora nobile Corcirese, Libri Ottavo, Al serenissimo principe ed eccellentiss.^{mo} Senato di Venetia., presso il Curti, Venetia, M.DC.LXXII/1672, pp. 399–417
26. Matton, R.: Corfou par Raymond Matton avec 139 photographies hors texte commentées en français et en anglais et 13 cartes et plans, pp.105, 182, fig. 109, Campanile de Potamia, Bibl. Gannadeion, 109/LXII
27. Mouzakis, C., Pantazis G.: Technical Report, Assessment of the seismic behavior of the Tower Bell Tower of St. Barbaros, Corfu (2019)
28. Moropoulou, A., et al.: Structural health monitoring of the holy aedicule in Jerusalem (2019)
29. Moropoulou, A., Cakmak, A.S., Biscontin, G., Zendri, E., et al.: Advanced Byzantine cement-based composites resisting earthquake stresses: the crushed brick/lime mortars of Justinian's Hagia Sophia, December 2002
30. Papageorgiou, S.K.: The History of the Church of Corfu. Broth. Aspioti G., Corfu, p. 230 (1920)
31. Partsch, J.: The island of Corfu: geographical monography, translation by Pericles Vegia, publ. I. Nachamouli, Corfu, pp.124–128 (1892)
32. Petit, L., Mgr, G.S.: Bibliographie des Akolouthies Grecques, Subsidia Hagiographica 16, Archevêque Latin d'Athènes, Bruxelles, pp. 22–23 (1926)
33. Steriotou, I.: The evolution of the" bastion system "in the fortifications of Greece, Friends of George P. Ekkekakis, "Antidorema", Honorary volume, Rethymno, pp. 277–296 (2013)

34. UNESCO/ICCROM/IUCN., World Heritage Resource Manual, Managing cultural world heritage (2013)
35. Papazachos, B., Papazachou, K.: The Earthquakes of Greece. Ziti Publications, Athens (2003). (in Greek)

Integrative Archaeological Surveying for the Enhancement of Nuraghe Piscu in Suelli (SU), Italy

Results and New Perspectives

Eleonora Scopinaro(✉), Roberto Gabrielli, and Andrea Angelini

Institute of Heritage Science, National Research Council, Rome, Italy
eleonora.scopinaro@ispc.cnr.it

Abstract. The research activity on Nuraghe Piscu represents the first step of a broader research project aimed at documentation, promotion, and enhancement of tangible and intangible cultural heritage of Suelli (SU), Italy.

The first excavations on this site date back to 1860 and further archaeological excavation and documentation campaigns were carried out in 1980s. However, present research shows the first complete survey of the monument. Thanks to this, it was finally possible to lay the foundations for an enhancement program that combines study and research with conservation and restoration activities using new technologies. The results obtained from the survey and analysis operations were useful both to deepen the research and documentation of the monument and to experiment with new methods of enhancement through 3D printing and augmented reality.

Keywords: Archaeological survey · 3D laser scanner · Texture mapping process · 3D printing · Enhancement · Nuraghe

1 Introduction

Suelli is a municipality located in the Texenta plain in central-southern Sardinia (Fig. 1). The territory is particularly rich in cultural heritage from the pre-historic period - Nuragic and pre-Nuragic - and subsequent eras due to the great strategic and cultural importance that the place has had over the millennia [1, 2]. Today, however, much of the local archaeological heritage is not accessible to the public and the town is not present in the major tourist routes despite its enormous potential.

The Nuraghe Piscu is one of the best preserved nuraghi in all of Sardinia. In addition to the central tower - which is preserved for almost its entire height - there are still four smaller towers joined by an enclosure with a large entrance in the south section. Many other structures are visible outside the enclosure, belonging to different construction phases and not yet fully investigated and excavated. Partially covered by the construction of the east tower there are the remains of another circular room probably belonging to a protonuraghe and around the enclosure along the south-east, south and west sides there are structures built later, probably attributable to the Punic, Roman and medieval eras [3, 4].

A. Osman and A. Moropoulou (Eds.): *Advanced Nondestructive and Structural Techniques for Diagnosis, Redesign and Health Monitoring for the Preservation of Cultural Heritage*, SPM 16, pp. 194–205, 2022.
https://doi.org/10.1007/978-3-031-03795-5_8

Fig. 1 Localization of Suelli, SU (Italy) on satellite image (http://www.google.it/maps - Imagery ©2021 TerraMetrics, Map data ©2021 Google, Inst. Geogr. Nacional) and a view of Nuraghe Piscu during surveying activities. Photo by R. Gabrielli (2018).

The first excavations of the Nuraghe Piscu were carried out in 1860 at the behest of Sir. Bartolomeo Casu, who at that time was the owner of the area. The results obtained were subsequently documented by Canon Giovanni Spano [5].

The first survey of the site was carried out only in the 1980s and consists of a 1:50 scale plan drawn up following further excavation campaigns, directed by Vincenzo Santoni, aimed at studying the monument and subsequent interventions of restoration [6].

It can therefore be state that the work carried out by the authors represents the first complete 3D survey of the Nuraghe Piscu in Suelli [7].

2 Documentation, Conservation, and Enhancement

The modern development of information technology and new technologies have opened new scenarios for the study, conservation, and enhancement of cultural heritage, especially in the field of archaeology, where the complexity and geographical and morphological limitations of some sites can make it difficult to access [8, 9]. Numerous difficulties are also encountered in the enjoyment of the finds, often removed from the places where they were found and not always accessible to the public due to problems of protection and/or space.

Project actions were based on the use of tools and techniques that can integrate the visit with mixed reality products, to reverse the paradigm of 'closure and inaccessibility,' which inevitably arises due to problems relating to the visitor's safety and protection of goods [10, 11].

2.1 The Project

The project for documentation, conservation, and enhancement of cultural heritage in Suelli aims at the recognition and preservation of tangible and intangible heritage as a promotion and support for the economic development of local communities, based on the Convention on the Value of Cultural Heritage for Society principals which identifies cultural heritage as "a group of resources inherited from the past which people identify" (Faro Convention 2005, art. 2). In fact, the conservation and enhancement of cultural heritage can have an important social function, if it is supported by systematic strategies aimed not only at cultural heritage, but also at all the other resources that characterize and represent all the distinctive historical signs of a region. The involvement of local communities, also implemented through the networking of the main stakeholders in the area, raises awareness of cultural heritage as the ability of citizens to recognize their identity in that heritage, to recognize it as their own and, consequently, to cooperate for its conservation.

The integrated providing of resources can also generate direct economic impacts, with the outsourcing of activities and services related to its management, as well as indirect impacts. The latter derive not only from the most well-known repercussions on the tourism industry, but also from the fact that the system that develops around the Heritage increases the competitiveness aspect of a territory, making it capable of attracting more than other human and financial resources, increasing tourist flows, as well as the establishment of productive activities not necessarily belonging to the cultural sector.

In order to obtain the best results in terms of enhancement, it is necessary to implement management policies to extend the use to places of culture that are currently less known and visited. The integration methods not only lead to a better knowledge of our territories but are also more sustainable from a tourist point of view; by relocating and distributing the flows of visitors over larger areas, on one hand the pressure on the sites normally visited is reduced, on the other hand the so-called 'minor' realities are enhanced, giving them greater economic sustainability.

The project aspires to strengthen the social sustainability of the tourism offer by increasing the use of cultural heritage through the study of enhancement models for cultural places on a territorial scale; the enhancement and systemization of the existing cultural heritage with new technologies; the planning of integrated territorial enhancement systems; the organization of educational projects for schools; the inclusion of the area in international scientific networks; the dissemination of the scientific results obtained and the inclusion of the area in the European and international tourism networks.

The enhancement of cultural heritage is based on the diversification of the visits and the increase in the socio-cultural development of local communities. To identify the most suitable tools for this territory, the strategies already used in other successful Italian and foreign projects were examined. Taking into account the characteristics of the place, in order to increase the accessibility and promotion of these sites, the following actions have been identified: analysis and evaluation of the strengths and criticalities of the tangible and intangible assets and the existing cultural offer; design of a network of fully accessible tourist routes that can be connected with other places of interest located in the neighboring municipalities; census, cataloguing and archiving in digital format of the existing cultural heritage; creation of three-dimensional models of objects and monuments through 3D laser scanner surveying and digital photogrammetry; scientific study and analysis of identified sites and objects; creation of multimedia contents to enhance the use of cultural heritage; creation of a web platform for the promotion and use of cultural heritage.

Thanks to the activation of high-level international partnerships, some already in place, it will be possible to raise, on an innovative level, the processes, and methodological approaches to be used in regional, national and community planning. These actions will favor the upward alignment of the competitive positioning of the participating companies, as well as the inclusion in national and international networks.

3 Methodology

Given the morphological features of the site and structures we decided to realize an instrumental survey with 3D laser scanner subsequently integrated with aerial photogrammetry from drone [12].

Data acquisition with laser scanner technology was obtained through the positioning of 92 stations, located at different heights, which allowed an overlap of the shooting areas always greater than 50%. The stations have been divided as follows: 5 inside the central tower; 3 on the top of the central tower; 32 inside the minor towers and in the courtyard; 14 above the structures of the secondary towers and the wall enclosure; 38 outside the wall enclosure (Fig. 2). The survey operations were carried out in seven consecutive days during the month of October 2018 with a phase-based laser scanner (FARO Focus 3D X330).

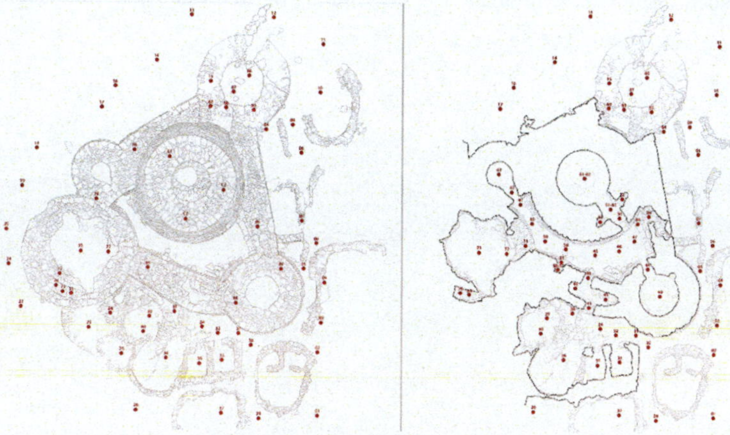

Fig. 2 Localization of laser scanner stations.

Subsequently, all the acquired material was recorded and processed through a dedicated software (Recontructor - Gexcel). The point clouds obtained from the different scans were 'cleaned' to eliminate any incorrect or too distant points and then they were aligned with each other through the recognition of homologous natural points (ICPs algorithm). Once the alignment was verified, the point clouds were then filtered and merged into a single unstructured cloud [13].

A further operation was the generation of a mesh based on the new point cloud; this made it possible to obtain surfaces that describe in detail the shape, measurements and materials of the artifact surveyed. This numerical model was used as a basis for the vectorization of the two-dimensional graphics (elevation plans and sections) and for 3D printing and projection [14].

At the same time spherical images were acquired with a 360° camera and aerial photogrammetric shots from drones [15]. Thanks to the photographic data it was possible to integrate the survey with HD images useful for creating a photorealistic texture on the virtual model. A specific protocol for texture mapping integration was developed considering the complexity of the structures and the organization of the spaces.

4 Results

The documentation activity was aimed at producing a multimedia model of the Nuraghe Piscu site, useful for representation in a virtual environment and for the creation of two-dimensional technical drawings which are essential for the design of restoration and enhancement intervention and for the planning of excavation activities (Fig. 3–4).

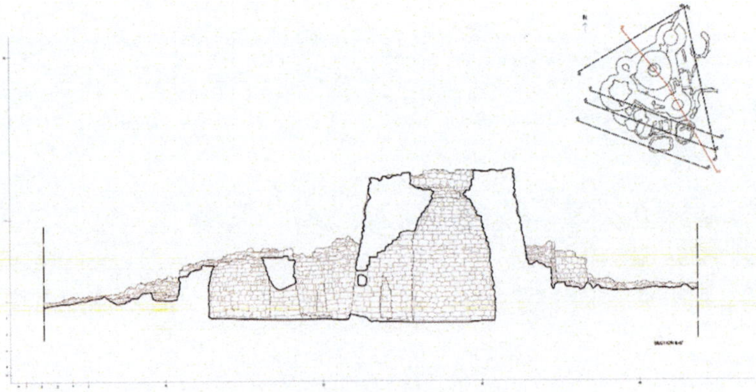

Fig. 3. Section of the main tower of Nuraghe Piscu.

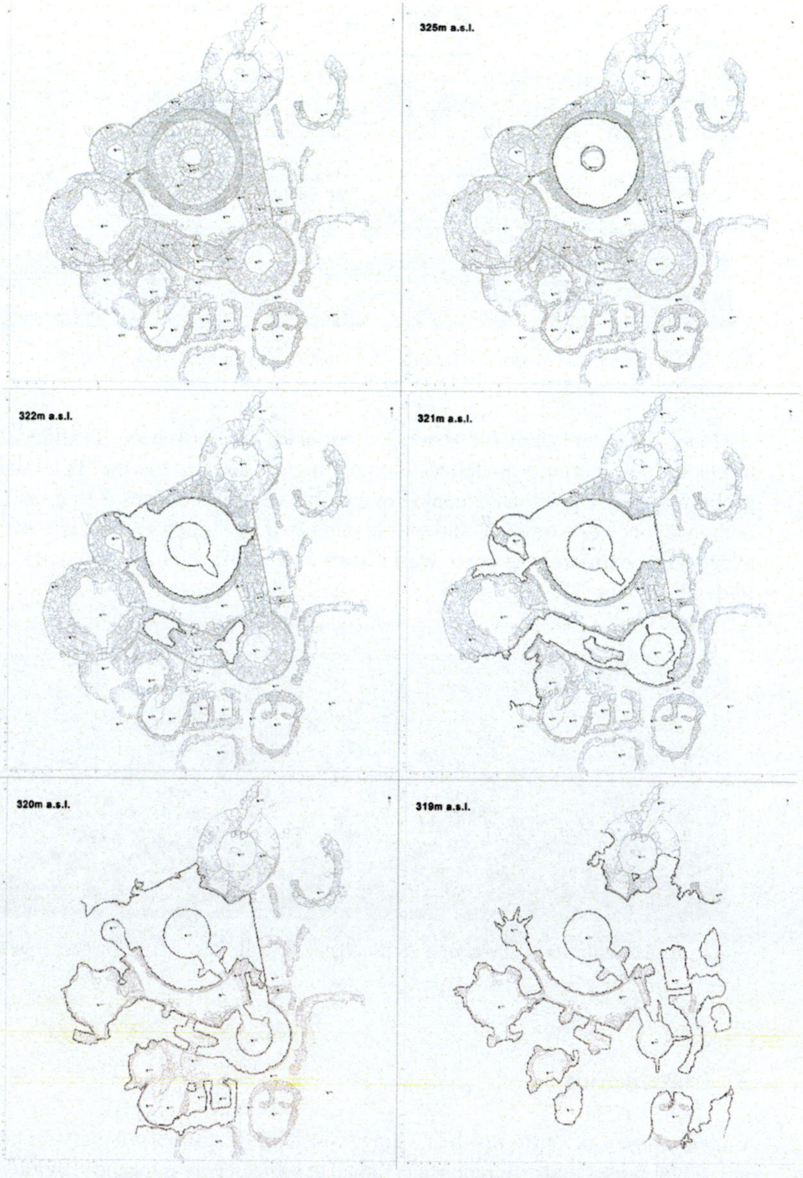

Fig. 4. Plan of the Nuraghe Piscu sectioned at different heights.

A three-dimensional print at 1:50 scale was also obtained by suitably optimizing the mesh of the numerical model [16]. This model has a surface treatment that simulates the stone material of the artefact itself and is usable for museum applications and as a basis for experimenting with the integration of tactile sensors as well (Fig. 5).

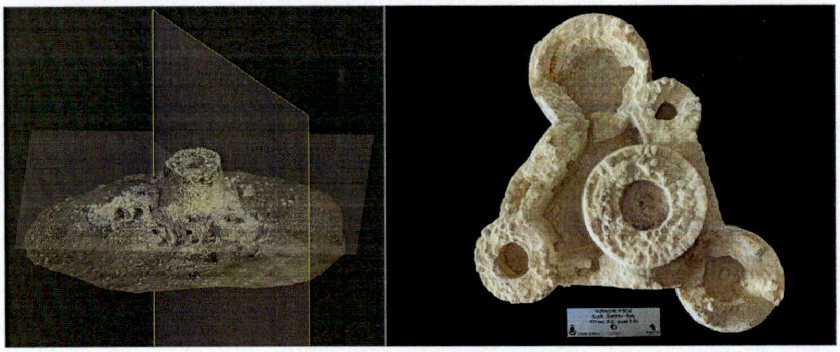

Fig. 5. The numerical model of Nuraghe Piscu (left side) and the 3D printed model (right side).

Most significant finds for exhibition purposes have also been identified. The same pieces have been photo-modelled to obtain high-resolution digital 3D models that are useful for study and enhancement (Fig. 6). One of the objectives is to transform visits, both in an open environment and within dedicated exhibition spaces, into multisensory experiences, enriching the use of works of art with additional information and interactive in-depth content [17, 18].

Fig. 6. The numerical model of a votive object virtually repositioned where it was found.

5 Conclusions

Built heritage is often the result of a long process of formation and transformation due to accidental causes and different needs linked to social, economic, and religious variations of the different historical-cultural contexts that they have gone through over time.

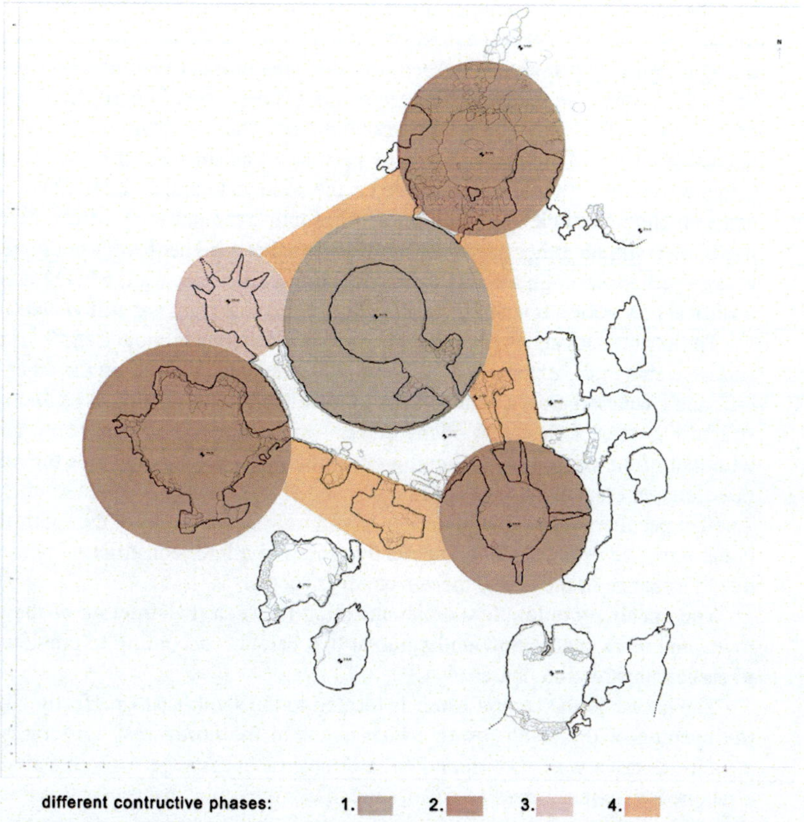

Fig. 7. Possible different construction phases identified by studying the different building techniques of the Nuraghe Piscu.

The Nuraghe Piscu archaeological site is not an exception. It is like an organism in continuous mutation and contains in its contemporary facies all the different moments that define its current consistency; from the construction phases to those of abandonment up to the rediscovery, to the past restoration interventions and to the new ones, that are nowadays in progress (Fig. 7).

By combining the study of the existing documentation with the results of the survey campaign, it is possible to draw some preliminary considerations on the constructive features of the different structures of the archaeological site of the Nuraghe Piscu and on the possible future prospects for research and enhancement.

The Nuraghe has a complex architecture characterized by the replication in plan of the 'tower' module. This type of construction involved the preliminary construction of a single central fortified tower (keep) to which the other structures were subsequently attached. In this case there are four secondary towers located - compared to the central one - to the north, south-east, south-west and west. The secondary towers are connected to each other and to the central tower by partitions that surround the complex forming a turreted bastion with a single entrance on the south side and an internal courtyard that connects the central element with the south, south-west and west towers. Access to the north tower, on the other hand, probably occurred at a higher level - no longer existing today - perhaps through a walkway inside the east enclosure wall, where the remains of a vertical connection open to the east sector of the courtyard are still visible.

The north, south and west towers have a plan size between about 6 and 9 m in diameter and have slits in the external parts. The south-west environment, on the other hand, has different characteristics, which perhaps followed different functions: the diameter of the structure is greater than that of the others - exceeding the size of 11 m, comparable with that of the central tower - and it does not have openings towards the outside, but three large internal niches and a staircase inserted in the masonry of the southern portion. Another peculiarity that emerged when examining the wall faces of the south-west tower is the presence of an opening apparently closed in a later construction phase, perhaps due to a change in the use of the surrounding rooms.

The opening would allow communication between the structure of the southwest tower and other spaces now underground that belong - according to previous studies - to earlier construction phases.

The access to the central tower is located to the south-east - perfectly aligned with the opening of the south tower, which opens to the north-west - in correspondence with the thickest wall portion and has a double niche passage without a stairwell. The structure is extremely massive with a solid masonry section that is over 4 m thick in some parts. The architecture of the nuragic chamber is well preserved for almost its potential entire height and shows a clear example of the corbelled construction technique with cantilevered blocks typical of this type of construction (Fig. 8).

Other structures are visible outside the enclosure, belonging to different construction phases and not yet fully investigated and excavated. Part of these structures are the subject of the excavation campaign in progress for the enhancement of the site, which will certainly bring new research perspectives for future projects.

However, it would be desirable not to limit the investigations to individual 'liberation' actions - although extremely important - to deepen the understanding of the entire site and all its construction phases. In fact, the elaboration of data of the land around the so far emerged architectures suggests the presence of many other structures that are still completely unknown. These structures could be the subject of new research aimed at increasing the diacritical reading of this archaeological site to protect its authenticity.

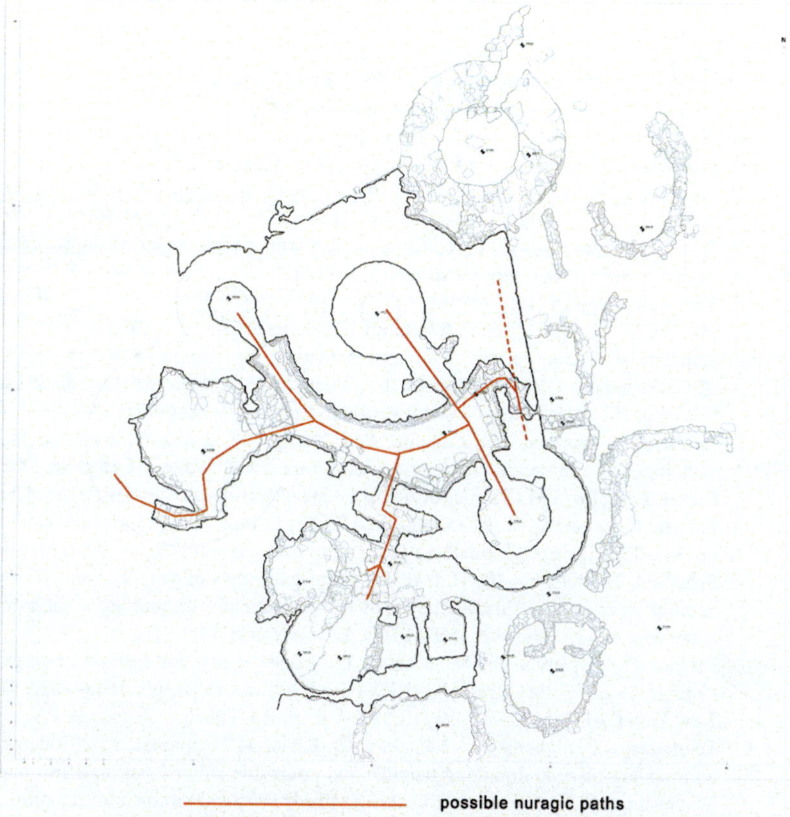

Fig. 8. Possible paths inside the complex during Nuragic period.

The results were used to enlarge accessibility and improve the use of the site, which will soon be finally reopened for tourist visits after decades of closure due to problems of danger and the lack of safe paths designed for everyone's needs.

The development of products for the enhancement of the site based on the results obtained is still in progress. The models are used for mixed reality applications in which the integration of 3D printing, AR and VR products will make it possible to implement the use of present multimedia results both in situ and remotely, in museums or even from home.

Acknoledgements. We thank the Municipality of Suelli and the Superintendence of Archeology, Fine Arts and Landscape for the metropolitan city of Cagliari and the provinces of Oristano and South Sardinia for having allowed and assisted the study operations. We also thank Teravista srl for the support during survey operations with drone.

References

1. Melis, P.: I nuraghi. In: Moravetti, A., Melis, P., Foddai, L., Alba, E. (eds.) La Sardegna nuragica. Storia e monumenti, pp. 28–53. Delfino, Sassari (2017)
2. Lilliu, G.: La civiltà nuragica, Delfino, Sassari (1982)
3. Melis, P.: Nuovi dati sull'architettura dei nuraghi a "tholos". In: Civiltà nuragica: nuove acquisizioni. Conference proceedings, pp. 27–43. Cagliari (2005)
4. Karimali, E.: Lithic technologies and use. In: Blake, E., Knapp, A. (eds.) The Archaeology of Mediterranean Prehistory, pp 180–214. Blackwell publishing Ltd, Hoboken (2004)
5. Spano, G.: Sopra i nuraghi di Sardegna. In: Bullettino archeologico sardo. Raccolta dei monumenti antichi in ogni genere di tutta l'isola di Sardegna, vol. 8, pp. 161–199. Tipografia Nazionale, Cagliari (1855–1856)
6. Santoni, V.: Nuraghe Piscu di Suelli: documenti materiali del bronzo medio-recente. In: La Sardegna nel Mediterraneo tra il bronzo medio e il bronzo recente (XVI-XII sec. a.C.). Atti del III Convegno di studi "Un millennio di relazioni tra la Sardegna e i Paesi del Mediterraneo", vol 3, pp. 167–185. Italian National Research Council, Rome (1991)
7. Gabrielli, R., Scopinaro, E., Angelini, A.: Il Nuraghe Piscu di Suelli tra archeologia e nuove tecnologie. Documentare. Conoscere. Valorizzare. Sandhi editore, Ortacesus (2020)
8. Correia Loureiro, S.M.: The use of augmented reality to expand the experience of museum visitors. In: Geroimenko, V. (ed.) Augmented Reality in Tourism, Museums and Heritage. SSCC, pp. 141–151. Springer, Heidelberg (2021). https://doi.org/10.1007/978-3-030-70198-7
9. Schofield, D., Johnson, T., Hufnal, D., Chapagain, P., Colletta, S., Lear, P.: Augmenting cultural experience: evaluating the use of augmented reality technology to enhance the visitor experience at a historic site. JSSSH **7**(2), 129–145 (2021)
10. Plecher, D.A., Wandinger, M., Klinker, G.: Mixed reality for cultural heritage. In: IEEE Conference on Virtual Reality and 3D User Interfaces (VR), pp. 1618–1622. IEEE Press, New York (2019). https://doi.org/10.1109/VR.2019.8797846
11. Doulamis, N., Doulamis, A., Ioannidis, C., Klein, M., Ioannides, M.: Modelling of static and moving objects: digitizing tangible and intangible cultural heritage. In: Ioannides, M., Thalmann, N.M., Papagiannakis, G. (eds.) Mixed reality and gamification for cultural heritage, pp 567–589. Springer, Cham (2017). https://doi.org/10.1007/978-3-319-49607-8_23
12. Remondino, F., Campana, S. (eds.) 3D Recording and Modelling in Archaeology and Cultural Heritage. Theory and best practice. BAR, Oxford (2014)
13. Bianchini, C., Inglese, C., Ippolito, A.: New analysis about archaeological architecture (AA): sic ancient theatres of the Mediterranean Sea. SCIRES **6**(2), 61–80 (2016). https://doi.org/10.2423/i22394303v6n2p61
14. Carpiceci, M., Carnevali, L., Angelini, A.: A new protocol for texture mapping process and 2d representation of rupestrian architecture. In: Zhalov, A., Gyorev, V., Delchev, P. (eds.) ISPRS TCII Symposium "Towards Photogrammetry 2020". International Archives of the Photogrammetry, Remote Sensing and Spatial Information Sciences, vol. 152, pp. 209–215 (2018)
15. Murtiyoso, A., Grussenmeyer, P.: Documentation of heritage buildings using Close-range UAV images: dense matching Issues, comparison, and case studies. Photogram. Record **32**(159), 206–229 (2017). https://doi.org/10.1111/phor.12197
16. Neumüller, N., Reichinger, A., Rist, F., Kern, C.: 3D printing for cultural heritage: preservation, accessibility, research and education. In: Ioannides, M., Quak, E. (eds.) 3D Research Challenges in Cultural Heritage. Lecture Notes in Computer Science, vol. 8355, pp. 119–134. Springer, Heidelberg (2014)

17. Doulamis, A., Doulamis, N., Protopapadakis, E., Voulodimos, A., Ioannides, M.: 4D modelling in cultural heritage. In: Ioannides, M., Martins, J., Žarnić, R., Lim, V. (eds.) Advances in Digital Cultural Heritage. Lecture Notes in Computer Science, vol. 10754, pp. 174–196. Springer, Cham (2018). https://doi.org/10.1007/978-3-319-75789-6_13
18. Doulamis, A., et al.: Selective 4D modelling framework for spatial-temporal land information management system. In: Third International Conference on Remote Sensing and Geoinformation of the Environment (RSCy2015), vol. 9535. International Society for Optics and Photonics (2015)

Author Index

A. Osman and A. Moropoulou (Eds.): *Advanced Nondestructive and Structural Techniques for Diagnosis, Redesign and Health Monitoring for the Preservation of Cultural Heritage*, SPM 16, pp. 207–208, 2022.
https://doi.org/10.1007/978-3-031-03795-5